1990

Basic Stochastic Processes
THE MARK KAC LECTURES

Basic Stochastic Processes

THE MARK KAC LECTURES

REZA IRANPOUR
DEPARTMENT OF MATHEMATICS
UNIVERSITY OF SOUTHERN CALIFORNIA

PAUL CHACON
DEPARTMENT OF MATHEMATICS
UNIVERSITY OF SOUTHERN CALIFORNIA

MACMILLAN PUBLISHING COMPANY
NEW YORK
COLLIER MACMILLAN PUBLISHERS
LONDON

Macmillan Publishing Company
866 Third Avenue, New York, New York 10022

Collier Macmillan Canada, Inc.

LIBRARY OF CONGRESS CATALOGING IN PUBLICATION DATA

Iranpour, Reza.
    Basic stochastic processes : the Mark Kac lectures / Reza Iranpour, Paul Chacon.
        p.   cm.
    Based on the lectures given by Mark Kac in a graduate course at the University of Southern California, Sept. 1982–Apr. 1983.
    1. Stochastic processes.   I. Chacon, Paul.   II. Kac, Mark.
III. Title.
QA274.I84 1988
519.2—dc19   87-19387
    ISBN 0-02-359820-4   CIP

Printing: 1 2 3 4 5 6 7 8   Year: 8 9 0 1 2 3 4 5 6 7

This book is dedicated to:

Mrs. Kitty Kac, and family
Mrs. Emily Miller, and family
and
our parents

Preface

This book is based on the lectures given by Professor Mark Kac in a graduate course at the University of Southern California from September, 1982 to April, 1983. The lectures were given to graduate students in engineering, mathematics and physics. The aim of the course was to introduce the basic stochastic processes using a minimum number of abstract mathematical concepts, in particular without reference to measure theory.

At the completion of his course, it became clear to Professor Kac that there was a need for a book on stochastic processes at this level. At his suggestion and under his constant supervision, Reza Iranpour, who was then his research assistant and had attended all the lectures, undertook the task of drafting such a text, based upon the class notes. Mark Kac and Reza Iranpour continued to work together very closely until Professor Kac's unfortunate and untimely death in November, 1984. Reza Iranpour then continued the work by himself, until he was joined by Paul Chacon in December, 1985.

The text is entirely self-contained, including a preliminary chapter, where the basics of probability theory are given, and an appendix consisting of the relevant background material in linear algebra. The main body of the text covers a wide range of stochastic processes, progressing naturally from the Poisson process to shot noise, Gaussian processes, Ornstein-Uhlenbeck processes, and Brownian motion, ending with a description of the Markov chains related to Brownian motion. Included are descriptions of Bochner's theorem, linear systems and the spectral density of stationary processes. Briefly mentioned are birth and death processes and renewal processes. The mathematical tools involved are restricted to standard Fourier series and Fourier transform techniques discussed in Chapters One and Two. Both the power and simplicity of this approach are evident throughout the text. In addition, we have attempted to preserve Mark Kac's lecture style, and to present his many devices to give the shortest and most elegant proofs.

Chapter Zero begins with the properties of random variables, and the moment generating function. Other topics in this Chapter include families of random variables, independence, the multivariate normal distribution and sums of independent random variables. The use of the moment generating function is stressed throughout, as a precursor to the use of the characteristic function later in the text. The Chapter ends with a section on the convergence of random variables, where we consider examples of the central limit theorem and the weak law of large numbers. Then we discuss the Borel-Cantelli lemmas and give a proof of the strong law, based on the Doob-Kolmogorov inequality.

Chapter One begins with the Fourier transform and a derivation of the Fourier inversion formula. Based on this we define the Dirac delta function and give some of its properties. These concepts are then put in a probabilistic framework when the characteristic function is introduced near the end of the chapter. Examples analogous to those solved by moment generating function methods are solved here using the characteristic function, and examples are given where the characteristic function can be used but the moment generating function cannot. The multidimensional analogues of these tools are introduced for later applications.

Chapter Two defines Fourier series as a mean square approximation to a square integrable function. Conditions for pointwise convergence of the series are stated and examples are given. The Fourier inversion formula is then shown to be a consequence of the convergence of Fourier series. Plancherel's formula is derived and used to show that the Fourier inversion formula holds for square integrable probability density functions. Later, by considering the Hilbert space of square integrable functions, we examine a geometric description of Fourier series. Finally, it is shown that positive definite functions have positive Fourier coefficients, which is a version of Bochner's theorem.

In Chapter Three we begin the study of stochastic processes with the Poisson process. This is presented first on an intuitive level, as a model for the number of emissions of a radioactive substance, then the formal definition of the Poisson process is given. Various equivalent descriptions of the Poisson process are then examined, showing that the exponential, uniform, Poisson random variables and the order statistics are connected with the Poisson process. Finally, generalizations of the Poisson process, including pure birth, birth and death, and renewal processes are briefly examined.

We introduce shot noise as a physical realization of the events in a Poisson process in Chapter Four. The idea is to determine what the output voltage of an instrument recording radioactive emissions should look like. The statistical moments of the resulting process are calculated. The concept of a linear system is introduced, and used to verify the calculations. Limiting cases of the shot noise process are shown to have interesting properties, and the shot noise example is used to introduce the concepts of stationarity, Gaussian processes and spectral density.

In Chapter Five the theory of Gaussian processes is presented. We begin with an informal discussion of the joint normal distribution. Some properties of the joint normal distribution are demonstrated and the historically significant rule for calculating higher correlations is derived. The characteristic function of a Gaussian process is worked out in detail. From this comes the covariance matrix, and it is shown that the mean and covariance of a Gaussian process completely determine the process. The Wiener process, stationary processes, and the spectral density of a stationary Gaussian process are considered. We also briefly examine the integral of a Gaussian process, and show that it is itself a Gaussian process.

In Chapter Six we introduce the Markov property and prove Doob's theorem, which says that the only stationary Markov Gaussian processes are the Ornstein-Uhlenbeck processes. This is done with the aid of the Chapman-Kolmogorov equation. As a generalization of Doob's theorem, it is shown how the conditional expectation of a Gaussian process can be used to determine whether or not the process is Markovian. We also examine the Hilbert space generated by a stochastic process, and using Bochner's theorem, sketch a proof of the spectral representation theorem for stationary processes.

With the groundwork now laid, in Chapter Seven a detailed analysis of Brownian motion is given, from a physical viewpoint. Formulas for the velocity and displacement of the Brownian particle are developed and their probability distributions are discussed. The Fokker-Planck differential equation is derived, and its connection to Brownian motion is presented. At the end is an analysis from the viewpoint of stochastic differential equations.

Chapter Eight begins with an introduction to the Markov chains which are related to Brownian motion, notably the Ehrenfest model and random walk. We show that by diagonalizing the transition matrix, one can obtain a great deal of information about these and other Markov chains. As an example of another specific type of Markov chain, and of the use of the probability generating function, we introduce branching processes. They are then used to discuss the question of extinction of family names. Later, in order to give some results applicable to general Markov chains, we show the relation between the stationary distribution and the mean recurrence time. The Chapter concludes with the ergodic properties of irreducible chains, and the convergence of the chain to the stationary distribution.

The Appendix Chapter briefly reviews the basic properties of matrices, systems of linear equations and linear transformations. Other short sections in this Chapter are square matrices, their structures, symmetric matrices, and quadratic forms. This review is mainly the linear algebra needed for the better understanding of Chapter Eight.

We would like to express our thanks to Professor G.C. Rota of M.I.T., for taking a significant amount of time out of his two months yearly stay at U.S.C. to consult on the project for the past few years. Professor M. Waterman of U.S.C. read the major portions of the book and gave numerous suggestions and criticisms. The constructive comments, help, and advice given by Dr. W.D. Miller on many parts of the project and earlier discussions with Dr. A. Chassiakos and Dr. S. Kharaghani have been very useful.

The final form of the book has been influenced by thoughtful reviews of Peter Ney, Mathematics Department, University of Wisconsin–Madison, and Richard Gundy, Department of Statistics, Rutgers University, New Brunswick, N.J.

We are very grateful for the enormous support, care and cooperation we have received from Professors R. Bruck (chairman), C. Lanski (undergraduate vice chairman), W. Proskurowski, A. Schumitzky, and A. Troesch of the Mathematics Department at U.S.C. They have been most generous, and helpful.

Thanks are also due to the former professors and present colleagues: S. Butler, M. Ho, L. C. Wellford, R. K. Miller, L. Redekopp, P. Seide, and F. Udwadia of the Engineering Department at U.S.C.; R. N. Blumenthal, B. Erickson, and R. Pyke of the Mathematics Department, University of Washington; R. Greene and J. White of the Mathematics Department at U.C.L.A.; S. Poffald of the Mathematics Department at Wabash College; and H. Kosaki of the College of General Education at Kyushu University, Japan for their excellent teaching, support and encouragement.

We owe many thanks to the science and mathematics editors, Mr. R. Clark, Mr. G. Ostedt and Mr. T. Maibach at Macmillan Publishing Company for their excellent service and communication. They have truly satisfied all of our wishes.

The service we have received from the staff of the Mathematics Department at U.S.C.: L. Burge-Ryerson, J. Carey, D. Lara, L. Lemons, and J. Mason throughout the project is invaluable.

Reza Iranpour

Paul Chacon

December, 1986

Contents

CHAPTER TWO

Fourier Series

CHAPTER THREE

Poisson Processes

CHAPTER FOUR

Shot Noise

CHAPTER FIVE
Gaussian Processes

CHAPTER SIX
Markov Gaussian Processes

CHAPTER SEVEN
Brownian Motion

CHAPTER EIGHT
Markov Chains

APPENDIX
Review of Linear Algebra

Index

CHAPTER

0

Preliminaries

Consider a random experiment: for example, the rolling of a die or the tossing of a dart at a target. The collection S of all possible outcomes of the experiment is called the **sample space** of the experiment. For the die it is a set of six elements, which are the six possible faces. For the dart it consists of all possible positions on the target. An **event** is a subset of the sample space. It could be one outcome, as in the event "a six is showing on the die"; or several outcomes, "a two or a three is showing"; or more complex, "the dart lands within 3 inches of the center of the target." The **probability** of an event E, which we shall write $P[E]$, is interpreted as the fraction of the time or "likelihood" that the outcome of the experiment is one of the outcomes in the set E. The following fundamental **rules of probability** are immediate consequences of this intuitive definition:

(i) $P[S] = 1$
(ii) $P[E] \geq 0$

(iii) $P\left[\bigcup_{i=1}^{\infty} E_i\right] = \sum_{i=1}^{\infty} P[E_i]$ if $E_i \cap E_j = \varnothing$ for $i \neq j$

Formally, P is a function defined on a class of subsets f of a set S, which satisfies (i), (ii), and (iii). For technical reasons it is not always possible for a probability to be assigned to all subsets of S, but if the collection of sets whose probability is defined is closed under countable unions, complements, and intersections (a "σ-field"), we say that the triple (S, f, P) is a **probability space**.

The **conditional probability** $P[A \mid B]$ of an event A, given an event B, is defined

1

by

$$P[A \mid B] = \frac{P[A \cap B]}{P[B]}, \qquad P[B] \neq 0 \tag{0.0.1}$$

This is interpreted as the likelihood that an outcome in B is also in A. The two events are said to be **independent** if being in B does not affect the chances of being in A, that is,

$$P[A \mid B] = P[A] \tag{0.0.2}$$

or in a more symmetric form,

$$P[A \cap B] = P[A]P[B] \tag{0.0.3}$$

A family of events E_1, E_2, . . . is said to be **independent** if for every finite subset E_{i_1}, E_{i_2}, . . . , E_{i_n} one has

$$P\left[\bigcap_{k=1}^{n} E_{i_k} \right] = \prod_{k=1}^{n} P[E_{i_k}] \tag{0.0.4}$$

It is worth remarking that even if there are only a finite number of events E_1, E_2, . . . , E_k, their independence cannot be assured merely by requiring that

$$P[E_1 \cap E_2 \cap \cdots \cap E_k] = P[E_1]P[E_2] \cdots P[E_k]$$

To demonstrate this, let S consist of the four points $S = \{1, 2, 3, 4\}$, with the probabilities

$$p_1 = \frac{\sqrt{2}}{2} - \frac{1}{4}, \qquad p_2 = \frac{1}{4}, \qquad p_3 = \frac{3}{4} - \frac{\sqrt{2}}{2}, \qquad p_4 = \frac{1}{4}$$

and consider the events

$$E_1 = \{1, 3\}, \qquad E_2 = \{2, 3\}, \qquad E_3 = \{3, 4\}$$

so that

$$E_1 \cap E_2 \cap E_3 = \{3\}, \qquad E_1 \cap E_2 = \{3\}$$

One can easily check that

$$P[E_1 \cap E_2 \cap E_3] = P[E_1]P[E_2]P[E_3]$$

but

$$P[E_1 \cap E_2] = p_3 \neq P[E_1]P[E_2]$$

A family of events $E_1, E_2, \ldots$ is said to be **pairwise independent** if

$$P[E_i \cap E_j] = P[E_i]P[E_j], \qquad i \neq j \tag{0.0.5}$$

Pairwise independence does not in general imply independence, as the following example shows. The sample space in this case consists of eight points,

$$S = \{1, 2, 3, 4, 5, 6, 7, 8\}$$

each with probability $\frac{1}{8}$. Let

$$E_1 = \{1, 2, 7, 8\}, \qquad E_2 = \{1, 4, 5, 8\}, \qquad E_3 = \{1, 3, 6, 8\}$$

and note that

$$P[E_1] = P[E_2] = P[E_3] = \frac{1}{2}$$

Moreover,

$$P[E_1 \cap E_2] = P[E_1]P[E_2]$$
$$P[E_1 \cap E_3] = P[E_1]P[E_3]$$
$$P[E_2 \cap E_3] = P[E_2]P[E_3]$$

but

$$P[E_1 \cap E_2 \cap E_3] \neq P[E_1]P[E_2]P[E_3]$$

Events $E_1, E_2, \ldots$ such that $E_i \cap E_j = \varnothing$ for $i \neq j$ are called **mutually exclusive.** Let $E_1, E_2, \ldots, E_n$ be a collection of mutually exclusive events whose union is S. The **total theorem of probability** states that

$$P[A] = \sum_{i=1}^{n} P[A \mid E_i]P[E_i] \tag{0.0.6}$$

and it follows immediately that

$$P[E_k \mid A] = \frac{P[E_k \cap A]}{P[A]} = \frac{P[A \mid E_k]P[E_k]}{\sum_{i=1}^{n} P[A \mid E_i]P[E_i]} \tag{0.0.7}$$

which is called **Bayes' theorem.**

0.1   RANDOM VARIABLES

There are many situations in which it is natural to consider functions of the outcome of a random experiment. Real-valued functions defined on a probability space are called **random variables,** that is, X is a random variable if for every $\xi \in S$, $X(\xi)$ is a real number.

Given a probability space (S, f, P) and a random variable X defined on S, the **distribution function** of X, denoted F_X, is defined by

$$F_X(x) = P[\{\xi \in S \text{ such that } X(\xi) \leq x\}] \tag{0.1.1}$$

or using the shorthand $\{X \leq x\} = \{\xi \in S \text{ such that } X(\xi) \leq x\}$,

$$F_X(x) = P[X \leq x] \tag{0.1.2}$$

The distribution function has the following properties:

(i) F_X is nondecreasing.
(ii) $0 \leq F_X \leq 1$
(iii) F_X is right continuous.
(iv) $\lim\limits_{x \to -\infty} F_X(x) = 0$, $\lim\limits_{x \to \infty} F_X(x) = 1$

Given a random variable X, if there exists an at most countable collection of points x_i and a nonnegative function p_X defined on these x_i such that

$$F_X(x) = \sum_{x_i \leq x} p_X(x_i) \tag{0.1.3}$$

then the random variable is said to be **discrete.** The distribution function of a discrete random variable grows by jumps, of size $p_X(x_i)$ at each x_i. The function p_X is called the **probability mass function** of the random variable, and

$$p_X(x_i) = P[X = x_i] \tag{0.1.4}$$

If $F_X(x)$ is differentiable, then X is called a **continuous random variable** and

$$f_X = \frac{dF_X}{dx} \tag{0.1.5}$$

is a nonnegative function, called the **density function** of X. This, by the fundamental theorem of the calculus, satisfies

$$F_X(x) = \int_{-\infty}^{x} f_X(t) \, dt \tag{0.1.6}$$

Further, any nonnegative function f_X for which

$$\int_{-\infty}^{\infty} f_X(t) \, dt = 1 \tag{0.1.7}$$

is called a probability density function, and the corresponding distribution function is found from (0.1.6). The **expected value,** denoted $E[X]$, of a discrete random variable X is defined by

$$E[X] = \sum_{x_i} x_i P[X = x_i] = \sum_{x_i} x_i p_X(x_i) \tag{0.1.8}$$

The expected value of a continuous random variable X is defined by

$$E[X] = \int_{-\infty}^{\infty} x f_X(x) \, dx \tag{0.1.9}$$

Given a discrete random variable X and a real-valued function h, the expected value of $h(X)$ can be found from

$$E[h(X)] = \sum_{x_i} h(x_i) P[X = x_i] = \sum_{x_i} h(x_i) p_X(x_i) \tag{0.1.10}$$

For a continuous random variable X the expected value of $h(X)$ can be calculated by

$$E[h(X)] = \int_{-\infty}^{\infty} h(x) f_X(x) \, dx \tag{0.1.11}$$

These are sometimes referred to as the law of the unconscious statistician. One can show that this is consistent with (0.1.8) and (0.1.9). Of particular interest is the case when we have

$$1_A(x) = \begin{cases} 1 & \text{if } x \in A \\ 0 & \text{else} \end{cases} \tag{0.1.12}$$

which is called the **indicator function** of the set A. For this function

$$E[1_A(X)] = P[X \in A]$$

Since indicator functions and sets are in a 1-to-1 correspondence and

$$1_{A \cap B} = 1_A 1_B$$

$$1_{A^c} = 1 - 1_A \tag{0.1.13}$$

$$1_{A \cup B} = 1_A + 1_B - 1_{A \cap B}$$

the indicator function can also be used to verify various formulas from set theory, for example, DeMorgan's formulas.

One way to define the expectation of a general random variable is to introduce the notion of a **Steiltjes integral.** We do this briefly here, leaving out some of the technical details. Suppose that F is a monotonically increasing function defined on $[a,b]$. Suppose P is a partition P of $[a,b]$:

$$P = \{x_0, \ldots, x_n\}$$

$$a = x_0 \leq x_1 \ldots \leq x_n = b$$

For any bounded real-valued function g defined on $[a,b]$, define the upper sum U by

$$U = \sum_{i=1}^{n} M_i(F(x_i) - F(x_{i-1}))$$

and the lower sum L by

$$L = \sum_{i=1}^{n} m_i(F(x_i) - F(x_{i-1})),$$

where

$$M_i = \sup \, (g(x)|x_{i-1} \leq x \leq x_i)$$

and

$$m_i = \inf \, (g(x)|x_{i-1} \leq x \leq x_i)$$

If over all partitions P

$$\inf_P U = \sup_P L$$

this value is called the Steiltjes integral of the function g with respect to the function F, and is written

$$\int_{[a,b]} g(x) \, dF(x)$$

The Riemann integral is a special case of this integral, where $F(x) = x$. The Steiltjes integral has similar properties to the Riemann integral; in particular, it is defined for any F and any continuous function defined on $[a,b]$. We stress that F need not be continuous. Integrals over open intervals (c,d) are defined as the limit of integrals over $[a,b]$ with a decreasing to c and b increasing to d. If F is not continuous, the integral over (a, b) and $[a, b]$ need not be equal, and we shall not use the notation

$$\int_a^b g(x) \, dF(x)$$

unless F is continuous. Integrals over the whole line are defined, as in the Riemann case, by the limit of integrals over $[a, b]$ as $a, b \to \pm\infty$. If the interval over which the integration takes place is omitted, as in

$$\int g(x) \, dF(x)$$

it is understood to be the whole real line.

The expectation of a random variable X can be defined as a Steiltjes integral of x with respect to its distribution function F_X, written

$$E[X] = \int x \, dF_X(x)$$

This formula applies if X is discrete or continuous, and can be taken to be the definition of the expected value of X, for any random variable X. We also have

$$E[g(X)] = \int g(x) \, dF_X(x)$$

The Steiltjes integral is a natural tool to use when we want to make statements or draw conclusions about general random variables. For making calculations, however, it is necessary to replace the integral by the corresponding sum if X is discrete, and by the integral against the density if X is continuous.

0.2 THE MOMENT GENERATING FUNCTION

We define the nth **moment** of a random variable to be the expectation $E[X^n]$, which may be calculated by setting $h(x) = x^n$ in (0.1.10) or (0.1.11). Related to the second moment are the central second moment, or **variance,** defined by

$$\text{var } X = E[(X - E[X])^2] = E[X^2] - (E[X])^2 \tag{0.2.1}$$

and the **standard deviation** σ_X, defined by

$$\sigma_X = \sqrt{\text{var } X} \tag{0.2.2}$$

For a random variable X, the following basic inequality is immediate:

$$P[X \geq a] \leq \frac{E[X]}{a} \qquad (0.2.3)$$

and by replacing X by $(X - E[X])^2$, we obtain **Chebyshev's inequality,**

$$P[|X - E[X]| \geq a] \leq \frac{\text{var } X}{a^2} \qquad (0.2.4)$$

A particularly useful random variable is the **moment generating function** M_X of a random variable X given by

$$M_X(t) = E[e^{tX}] \qquad (0.2.5)$$

If two random variables have the same moment generating function, they have the same distribution function. We shall use this fact to determine the distribution function of some random variables defined as combinations of known random variables in Section (0.3). By Taylor series we can write the moment generating function in the form

$$M_X(t) = 1 + E[X]t + E[X^2]\frac{t^2}{2!} + E[X^3]\frac{t^3}{3!} + \cdots \qquad (0.2.6)$$

which gives the useful formula

$$E[X^n] = \frac{d^n M_X}{dt^n}\bigg|_{t=0} \qquad (0.2.7)$$

The coefficients K_i in the expansion

$$\ln M_X(t) = K_1 t + K_2 \frac{t^2}{2!} + K_3 \frac{t^3}{3!} + \cdots \qquad (0.2.8)$$

are called the **cumulants** of the random variable X. Since $K_2, K_3, \ldots$ are invariant under translations of the random variable (i.e., they have the same values for X and $X + a$), they are also called the semi-invariants of X. By (0.2.6),

$$\ln M_X(t) = \ln\left(1 + E[X]t + E[X^2]\frac{t^2}{2!} + E[X^3]\frac{t^3}{3!} + \cdots\right)$$

which we can expand using

$$\ln(1 + x) = x - \frac{x^2}{2} + \frac{x^3}{3} - \frac{x^4}{4} + \cdots$$

and compare with (0.2.8) to get the values of the cumulants in terms of the moments,

$$K_1 = E[X]$$
$$K_2 = E[X^2] - (E[X])^2 \qquad\qquad (0.2.9)$$
$$K_3 = E[X^3] - 3E[X]E[X^2] + 2(E[X])^3$$

$$\vdots$$

We note that K_1 and K_2 are the mean and variance of X.

0.3   DISCRETE RANDOM VARIABLES

Bernoulli trials

A sequence of random variables $X_1, X_2, \ldots$ is called a sequence of **Bernoulli trials** if each random variable takes on only the values 0 and 1, with

$$P[X_i = 1] = p, \qquad P[X_i = 0] = 1 - p, \qquad 0 \le p \le 1$$

and if the events $\{X_i = 1, i = 1, 2, \ldots\}$ are an independent family. A familiar example is coin tossing, if we assign the value 1 to "heads" and 0 to "tails," which has $p = \frac{1}{2}$. In this section two important discrete random variables are introduced which are related to Bernoulli trials. We calculate their moment generating functions, means, and variances.

Binomial random variables

The **binomial random variable** $B_{n,p}$ counts the number of "successes" obtained in n Bernoulli trials and is given by

$$B_{n,p} = \sum_{k=1}^{n} X_k \qquad\qquad (0.3.1)$$

with X_i as above. This random variable has mass function

$$P[B_{n,p} = k] = \binom{n}{k} p^k (1 - p)^{n-k} \qquad\qquad (0.3.2)$$

and is denoted by B_n if $p = \frac{1}{2}$,

$$P[B_n = k] = \binom{n}{k} (\tfrac{1}{2})^n$$

This indeed defines a probability mass function, since it is immediate from the binomial theorem that

$$\sum_{k=0}^{n} \binom{n}{k} p^k (1 - p)^{n-k} = [p + (1 - p)]^n = 1$$

We find the expectation of $B_{n,p}$,

$$
\begin{aligned}
E[B_{n,p}] &= \sum_{k=0}^{n} k \binom{n}{k} p^k (1 - p)^{n-k} \\
&= np \sum_{k=1}^{n} \binom{n-1}{k-1} p^{k-1} (1 - p)^{n-k} = np
\end{aligned}
$$
(0.3.3)

and its second moment

$$E[(B_{n,p})^2] = \sum_{k=0}^{n} k^2 \binom{n}{k} p^k (1 - p)^{n-k}$$
(0.3.4)

which since $k^2 = k(k - 1) + k$,

$$= n(n + 1)p^2 \sum_{k=2}^{n} \binom{n-2}{k-2} p^{k-2} (1 - p)^{n-k} + np \sum_{k=1}^{n} \binom{n-1}{k-1} p^{k-1} (1 - p)^{n-k}$$

$$= n(n - 1)p^2 + np$$

From these the variance of $B_{n,p}$ is

$$E[(B_{n,p})^2] - (E[B_{n,p}])^2 = np(1 - p)$$
(0.3.5)

We also find its moment generating function,

$$
\begin{aligned}
M_{B_{n,p}}(t) &= E[\exp (tB_{n,p})] \\
&= \sum_{k=0}^{n} e^{tk} \binom{n}{k} p^k (1 - p)^{n-k} \\
&= [pe^t + (1 - p)]^n
\end{aligned}
$$
(0.3.6)

and use it to verify the first moment

$$E[B_{n,p}] = \frac{dM_X(t)}{dt}\bigg|_{t=0} = np$$

and the second moment

$$E[(B_{n,p})^2] = \left.\frac{d^2M_X(t)}{dt^2}\right|_{t=0} = n(n-1)p^2 + np$$

Poisson random variables

The Poisson distribution arises as an approximation to the binomial distribution. If n is very large and p very small and the product $\lambda = np$ is of moderate size, a binomial random variable $B_{n,p}$ has a mass function which is approximately the same as the mass function of the Poisson random variable N_λ defined by

$$P[N_\lambda = k] = \frac{e^{-\lambda}\lambda^k}{k!}, \qquad k = 0, 1, 2, \ldots \qquad (0.3.7)$$

This can be shown using **Stirling's formula,**

$$n! \sim \sqrt{2\pi n}\, n^n e^{-n} \qquad (0.3.8)$$

where $\sim$ means that the ratio of the two quantities approaches 1 as $n \to \infty$. For large n, this formula is one way to approximate mass functions which involve factorials and we use it in Section 0.8. It is left to the reader to show, under the conditions mentioned above, that in the binomial case this substitution results in the Poisson distribution.

We calculate the expectation of N_λ,

$$\begin{aligned}
E[N_\lambda] &= \sum_{k=0}^{\infty} k\,\frac{e^{-\lambda}\lambda^k}{k!} \\
&= \lambda \sum_{k=1}^{\infty} \frac{e^{-\lambda}\lambda^{k-1}}{(k-1)!} = \lambda
\end{aligned} \qquad (0.3.9)$$

and its second moment, using $k^2 = k(k-1) + k$,

$$\begin{aligned}
E[N_\lambda^2] &= \sum_{k=0}^{\infty} k^2\,\frac{e^{-\lambda}\lambda^k}{k!} \\
&= \sum_{k=2}^{\infty} \frac{e^{-\lambda}\lambda^k}{(k-2)!} + \sum_{k=1}^{\infty} \frac{e^{-\lambda}\lambda^k}{(k-1)!} \\
&= \lambda^2 + \lambda
\end{aligned} \qquad (0.3.10)$$

so that the variance of N_λ is

$$E[N_\lambda^2] - (E[N_\lambda])^2 = \lambda \qquad (0.3.11)$$

Its moment generating function is

$$M_{N_\lambda}(t) = E[\exp(tN_\lambda)]$$

$$= \sum_{k=0}^{\infty} e^{tk} \frac{e^{-\lambda}\lambda^k}{k!} = \exp(\lambda e^t - \lambda)$$

(0.3.12)

from which we can verify

$$E[N_\lambda] = \left.\frac{dM_X(t)}{dt}\right|_{t=0} = \lambda$$

and

$$E[N_\lambda^2] = \left.\frac{d^2M_X(t)}{dt^2}\right|_{t=0} = \lambda^2 + \lambda$$

One can check that all the cumulants of N_λ are equal to λ, since

$$\ln M_{N_\lambda}(t) = \lambda(e^t - 1) = \lambda\left(t + \frac{t^2}{2!} + \frac{t^3}{3!} + \cdots\right)$$

0.4   CONTINUOUS RANDOM VARIABLES

Normal random variables

Perhaps the most important continuous random variables are those whose density is of the form

$$f_X(x) = \frac{1}{\sigma\sqrt{2\pi}} \exp\left[-\frac{1}{2}\left(\frac{x-\mu}{\sigma}\right)^2\right]$$

(0.4.1)

Such a random variable is called a **normal** random variable of mean μ and variance σ^2 and is denoted $N(\mu, \sigma^2)$. To verify that this formula defines a density, we first check that

$$\int_{-\infty}^{\infty} \exp(-x^2)\, dx = \sqrt{\pi}$$

(0.4.2)

which is clear from

$$\left[\int_{-\infty}^{\infty} \exp(-x^2)\, dx\right]^2 = \int_{-\infty}^{\infty} \exp(-x^2)\, dx \int_{-\infty}^{\infty} \exp(-y^2)\, dy$$

$$= \int_{-\infty}^{\infty}\int_{-\infty}^{\infty} \exp[-(x^2+y^2)]\, dx\, dy$$

or in polar coordinates,

$$= \int_0^{2\pi}\int_0^{\infty} \exp(-r^2) r\, dr\, d\theta$$

$$= \pi$$

By a change of variables it is then immediate that

$$\int_{-\infty}^{\infty} \frac{1}{\sigma\sqrt{2\pi}} \exp\left[-\frac{1}{2}\left(\frac{x-\mu}{\sigma}\right)^2\right] = 1$$

One can show directly that μ and σ^2 are the mean and variance, but we shall apply the moment generating function:

$$M_X(t) = \int_0^{\infty} e^{tx} f_X(x)\, dx$$

$$= \int_0^{\infty} e^{tx} \frac{1}{\sigma\sqrt{2\pi}} \exp\left[-\frac{1}{2}\left(\frac{x-\mu}{\sigma}\right)^2\right] dx$$

(0.4.3)

which by completing the square of the exponent and using (0.4.2) becomes

$$= e^{\mu t + (1/2)\sigma^2 t^2}$$

Thus one finds the mean,

$$\left.\frac{dM_X(t)}{dt}\right|_{t=0} = \mu$$

and the second moment,

$$\left.\frac{d^2 M_X(t)}{dt^2}\right|_{t=0} = \sigma^2 + \mu^2$$

which give the variance,

$$E[X^2] - (E[X])^2 = (\sigma^2 + \mu^2) - \mu^2 = \sigma^2$$

We note that all the rest of the cumulants are zero, since

$$\ln M_X(t) = \mu t + \tfrac{1}{2}\sigma^2 t^2$$

Cauchy random variables

A **Cauchy** random variable X with parameter θ has the density function

$$f_X(x) = \frac{1}{\pi} \frac{1}{1 + (x - \theta)^2}, \tag{0.4.4}$$

It is left as an exercise to show that all the even moments of this random variable are infinite and that the moment generating function for this random variable does not exist for any value of t.

Uniform random variables

A random variable X which is equally likely to assume any value in an interval $[a, b]$ is called **uniform.** The corresponding density function is

$$f_X(x) = \begin{cases} \dfrac{1}{b - a}, & a \le x \le b \\ 0, & \text{else} \end{cases} \tag{0.4.5}$$

The first and second moments are

$$E[X] = \frac{1}{b - a} \int_a^b x \, dx = \frac{b - a}{2} \tag{0.4.6}$$

and

$$E[X^2] = \frac{1}{b - a} \int_a^b x^2 \, dx = \frac{b^2 + ab + a^2}{3} \tag{0.4.7}$$

so that

$$\operatorname{var} X = \frac{(b - a)^2}{12} \tag{0.4.8}$$

We also compute the moment generating function,

$$M_X(x) = \frac{1}{b - a} \int_a^b e^{tx} \, dx = \frac{e^{tb} - e^{ta}}{t(b - a)} \tag{0.4.9}$$

which could be used to verify the first and second moments.

Exponential random variables

If a random variable X has density

$$f_X(x) = \begin{cases} \lambda e^{-\lambda x}, & x \geq 0 \\ 0, & x < 0 \end{cases} \tag{0.4.10}$$

then X is said to be exponential with parameter λ. Its moments may be computed by integration by parts, but we find them by calculating the moment generating function:

$$M_X(t) = \int_0^\infty e^{tx} \lambda e^{-\lambda x} \, dx$$

$$= \frac{\lambda}{\lambda - t} \tag{0.4.11}$$

$$= 1 + \frac{t}{\lambda} + \left(\frac{t}{\lambda}\right)^2 + \left(\frac{t}{\lambda}\right)^3 + \cdots$$

so that the nth moment is

$$E[X^n] = \frac{n!}{\lambda^n} \tag{0.4.12}$$

and from this

$$\text{var } X = \frac{2}{\lambda^2} - \frac{1}{\lambda^2} = \frac{1}{\lambda^2} \tag{0.4.13}$$

Since

$$\ln M_X(t) = -\ln\left(1 - \frac{t}{\lambda}\right)$$

$$= \frac{t}{\lambda} + \frac{1}{2}\left(\frac{t}{\lambda}\right)^2 + \frac{1}{3}\left(\frac{t}{\lambda}\right)^3 + \cdots$$

the general formula for the cumulants is

$$K_n = \frac{(n-1)!}{\lambda^n} \tag{0.4.14}$$

0.5 FAMILIES OF RANDOM VARIABLES, INDEPENDENCE

If more than one random variable is defined on a probability space, the random variables are said to be **jointly distributed.** In addition to each random variable having a distribution function, there is a distribution function for the collection as a whole.

This is called the **joint distribution function** of the family of random variables. If $X_1, \ldots, X_n$ is a family of jointly distributed random variables, the joint distribution function of $\mathbf{X} = (X_1, \ldots, X_n)$, denoted $F_\mathbf{X}$, is defined by

$$F_\mathbf{X}(x_1, \ldots, x_n) = P[X_1 \leq x_1, \ldots, X_n \leq x_n] \qquad (0.5.1)$$

and has the following properties:

(i) $F_\mathbf{X}(x_1, \ldots, x_n)$ is nondecreasing in any x_i.
(ii) $F_\mathbf{X}(x_1, \ldots, x_n)$ tends to zero as any x_i tends to $-\infty$.
(iii) $F_\mathbf{X}(x_1, \ldots, x_n)$ tends to one as all x_i tend to ∞.
(iv) $0 \leq F_\mathbf{X}(x_1, \ldots, x_n) \leq 1$
(v) $F_\mathbf{X}(x_1, \ldots, x_n)$ is right continuous in any x_i.

If there exists a nonnegative function $f_\mathbf{X}$ such that

$$F_\mathbf{X}(x_1, \ldots, x_n) = \int_{-\infty}^{x_n} \cdots \int_{-\infty}^{x_1} f_\mathbf{X}(y_1, \ldots, y_n) \, dy_1 \cdots dy_n \quad (0.5.2)$$

then the random variables are said to be **jointly continuous,** and $f_\mathbf{X}$ is called the **joint density function** of $\mathbf{X}$. By the fundamental theorem of the calculus, if $F_\mathbf{X}$ is differentiable, the random variables are jointly continuous and their joint density is

$$f_\mathbf{X}(x_1, \ldots, x_n) = \frac{\partial^n F_\mathbf{X}(x_1, \ldots, x_n)}{\partial x_1 \cdots \partial x_n} \qquad (0.5.3)$$

For discrete random variables the **joint mass function** is

$$p_\mathbf{X}(x_1, \ldots, x_n) = P[X_1 = x_1, \ldots, X_n = x_n] \qquad (0.5.4)$$

An important property of the joint density function is that the joint density for a smaller collection of random variables can be found by "integrating out" the unwanted variables. For example, if $\mathbf{Y} = (x_2, \ldots, x_n)$, then the joint density for $\mathbf{Y}$ is

$$f_\mathbf{Y}(x_2, \ldots, x_n) = \int_{-\infty}^{\infty} f_\mathbf{X}(x_1, \ldots, x_n) \, dx_1$$

For discrete random variables, we have

$$p_\mathbf{Y}(x_2, \ldots, x_n) = \sum_{x_1} p_\mathbf{X}(x_1, \ldots, x_n)$$

For two random variables X and Y, expectation has the basic property

$$E[X + Y] = E[X] + E[Y]$$

If X and Y are continuous, by applying the law of the unconscious statistician, we can now verify this property:

$$E[X + Y] = \int_{-\infty}^{\infty} \int_{-\infty}^{\infty} (x + y) f_{X,Y}(x, y) \, dx \, dy$$

$$= \int_{-\infty}^{\infty} \int_{-\infty}^{\infty} x f_{X,Y}(x, y) \, dx \, dy + \int_{-\infty}^{\infty} \int_{-\infty}^{\infty} y f_{X,Y}(x, y) \, dx \, dy$$

$$= \int_{-\infty}^{\infty} x f_X(x) \, dx + \int_{-\infty}^{\infty} y f_Y(y) \, dy$$

$$= E[X] + E[Y]$$

The proof in the discrete case is similar. One can also show that if $X_1, \ldots, X_n$ is a family of random variables, then

$$E[X_1 + \cdots + X_n] = E[X_1] + \cdots + E[X_n]$$

The **joint moment generating function** of $\mathbf{X} = (X_1, \ldots, X_n)$, where the X_i's are jointly distributed random variables, is defined by

$$M_{\mathbf{X}}(t_1, \ldots, t_n) = E\left[\exp\left(\sum_{i=1}^{n} t_i X_i \right) \right] \tag{0.5.5}$$

Suppose that X and Y have joint density $f_{X,Y}$. The **conditional distribution** function $F_{X|Y}(x|y)$ of X given Y is found by considering

$$\lim_{\epsilon \to 0} P[X \le x | y < Y \le y + \epsilon]$$

which yields, assuming that $f_{X,Y}$ is continuous and f_Y is strictly positive,

$$F_{X|Y}(x|y) = \frac{1}{f_Y(y)} \int_{-\infty}^{x} f_{X,Y}(x, y) \, dx \tag{0.5.6}$$

The corresponding **conditional density** is

$$f_{X|Y}(x|y) = \frac{1}{f_Y(y)} f_{X,Y}(x, y) \tag{0.5.7}$$

When conditioning on more than one variable, we get

$$f_{X|Y_1,\ldots,Y_n}(x|y_1, \ldots, y_n) = \frac{1}{f_{Y_1,\ldots,Y_n}(y_1, \ldots, y_n)} f_{X,Y_1,\ldots,Y_n}(x, y_1, \ldots, y_n)$$

For discrete random variables X and Y, the **conditional mass function** is

$$p_{X|Y}(x_i|y) = P[X = x_i|Y = y] = \frac{p_{X,Y}(x_i, y)}{p_Y(y)} \qquad (0.5.8)$$

One can easily check that the conditional density is, in fact, a density function,

$$\int_{-\infty}^{\infty} f_{X|Y}(x|y)\, dx = 1$$

and that the conditional mass function is a mass function

$$\sum_{i=1}^{\infty} p_{X|Y}(x_i|y) = 1$$

Corresponding to the conditional distribution is the **conditional expectation** of a random variable X with respect to a random variable Y, which can be thought of as the average value that X has when the value of Y is known. For discrete random variables this is

$$E[X|Y = y] = \sum_{x_i} x_i p_{X|Y}(x_i|y)$$

and for continuous random variables

$$E[X|Y = y] = \int_{-\infty}^{\infty} x f_{X|Y}(x|y)\, dx$$

There are two interpretations of conditional expectation: one, as above, where the value of Y is known, and another, where it is considered as a function of the random variable Y and thus a random variable itself. In the latter case it is denoted as

$$E[X|Y]$$

We note two important properties of conditional expectation:

$$E[E[X|Y]] = E[X]$$

and

$$E[YE[X|Y]] = E[XY]$$

Conditional expectation is a type of projection that can be used to make random variables X and Y **orthogonal,** by which we mean

$$E[XY] = 0$$

Consider two jointly distributed random variables X and Y. Subtracting the conditional expectation $E[X|Y]$ from X, we get a random variable orthogonal to Y, for if

$$Z = X - E[X|Y]$$

then

$$
\begin{aligned}
E[ZY] &= E[(X - E[X|Y])(Y)] \\
&= E[XY] - E[YE[X|Y]] \\
&= E[XY] - E[XY] \\
&= 0
\end{aligned}
$$

Two jointly distributed random variables X and Y are said to be **independent** if

$$P[X \le x, \, Y \le y] = P[X \le x]P[Y \le y] \tag{0.5.9}$$

or

$$F_{X,Y}(x, \, y) = F_X(x)F_Y(y) \tag{0.5.10}$$

If X and Y have density functions f_X, f_Y, from this we get

$$
\begin{aligned}
\int_{-\infty}^{x} \int_{-\infty}^{y} f_{X,Y}(t, \, s) \, dt \, ds &= \int_{-\infty}^{x} f_X(t) \, dt \int_{-\infty}^{y} f_Y(s) \, ds \\
&= \int_{-\infty}^{x} \int_{-\infty}^{y} f_X(t)f_Y(s) \, dt \, ds
\end{aligned}
$$

or

$$f_{X,Y}(t, \, s) = f_X(t)f_Y(s) \tag{0.5.11}$$

An important consequence of the independence of two random variables X and Y is that they are **uncorrelated,** that is,

$$E[XY] = E[X]E[Y] \tag{0.5.12}$$

It is clear that if X and Y are independent random variables, then e^{tX} and e^{sY} are independent, and in this case,

$$E[e^{tX}e^{sY}] = E[e^{tX}]E[e^{sY}]$$

which is the same as

$$M_{X,Y}(t, s) = M_X(t)M_Y(s) \tag{0.5.13}$$

The **covariance** of two random variables X and Y is defined by

$$\text{cov}(X, Y) = E[(X - E[X])(Y - E[Y])] \tag{0.5.14}$$

which if X and Y are independent becomes

$$\text{cov}(X, Y) = 0$$

Related to the covariance is the **Cauchy–Schwarz inequality,**

$$(E[XY])^2 \leq E[X^2]E[Y^2] \tag{0.5.15}$$

which implies that

$$[\text{cov}(X, Y)]^2 \leq (\sigma_X)^2(\sigma_Y)^2 \tag{0.5.16}$$

We define the **correlation coefficient** $\rho_{X,Y}$ by

$$\rho_{X,Y} = \frac{\text{cov}(X_i, X_j)}{\sigma_X \sigma_Y} \tag{0.5.17}$$

which by (0.5.16) has the property

$$|\rho_{X,Y}| \leq 1$$

A jointly distributed family of random variables $X_1, X_2, \ldots, X_n$ is said to be **pairwise independent** if

$$P[X_i \leq x_i, X_j \leq x_j] = P[X_i \leq x_i]P[X_j \leq x_j], \qquad i \neq j \tag{0.5.18}$$

or

$$F_{X_i, X_j}(x_i, x_j) = F_{X_i}(x_i)F_{X_j}(x_j), \qquad i \neq j$$

and **mutually independent** if

$$F_X(x_1, \ldots, x_n) = F_{X_1}(x_1) \cdots F_{X_n}(x_n)$$

which implies that

$$P[X_{i_1} \leq x_{i_1}, X_{i_2} \leq x_{i_2}, \ldots, X_{i_k} \leq x_{i_k}] = P[X_{i_1} \leq x_{i_1}]P[X_{i_2} \leq x_{i_2}] \cdots P[X_{i_k} \leq x_{i_k}]$$
(0.5.19)

for any collection of indices i_k where no index is repeated. If the random variables are jointly continuous, their mutual independence is equivalent to the joint density function factoring into the product of the individual density functions. If the joint moment generating function exists, mutual independence is equivalent to

$$M_\mathbf{X}(t_1, \ldots, t_n) = \prod_{k=1}^{n} M_{X_k}(t_k)$$
(0.5.20)

We say that a family of random variables $(X_1, \ldots, X_m)$ is independent of a family of random variables $(Y_1, \ldots, Y_n)$, if for all real $x_1, \ldots, x_m$ and $y_1, \ldots, y_n$, we have

$$P[X_1 \leq x_1, \ldots, X_m \leq x_m, Y_1 \leq y_1, \ldots, Y_n \leq y_n]$$
$$= P[X_1 \leq x_1, \ldots, X_m \leq x_m]P[Y_1 \leq y_1, \ldots, Y_n \leq y_n]$$
(0.5.21)

If the joint moment generating functions exists, this is equivalent to

$$M_{\mathbf{X},\mathbf{Y}}(t_1, \ldots, t_m, s_1, \ldots, s_n) = M_\mathbf{X}(t_1, \ldots, t_m)M_\mathbf{Y}(s_1, \ldots, s_n)$$
(0.5.22)

0.6 MULTIVARIATE NORMAL DISTRIBUTION

A family of random variables $X_1, \ldots, X_n$ is said to be jointly **normally distributed** if their joint density function is

$$\begin{aligned}
f(x_1, \ldots, x_n) &= \frac{\sqrt{\det \mathbf{A}}}{\sqrt{(2\pi)^n}} \exp\left[-\tfrac{1}{2} \sum_{i=1}^{n} \sum_{j=1}^{n} a_{ij}(x_i - \mu_i)(x_j - \mu_j)\right] \\
&= \frac{\sqrt{\det \mathbf{A}}}{\sqrt{(2\pi)^n}} \exp\left[-\tfrac{1}{2}(\mathbf{x} - \mathbf{\mu})^\mathsf{T}\mathbf{A}(\mathbf{x} - \mathbf{\mu})\right]
\end{aligned}$$
(0.6.1)

where $\mathbf{A} = (a_{ij})$ is a symmetric positive definite matrix and $\mathbf{\mu}^\mathsf{T} = (\mu_1, \ldots, \mu_n)$. Here each X_i is normally distributed with

$$E[X_i] = \mu_i$$
(0.6.2)

and

$$\text{cov}(X_i, X_j) = \mathbf{A}^{-1}$$
(0.6.3)

The joint moment generating function of $\mathbf{X}$ is

$$E\left[\exp\left(\sum_{i=1}^{n} t_i X_i\right)\right] = \exp\left(\sum_{i=1}^{n} \mu_i t_i + \tfrac{1}{2}\sum_{i=1}^{n}\sum_{j=1}^{n} r_{ij} t_i t_j\right) \qquad (0.6.4)$$

where

$$(r_{ij}) = \mathbf{A}^{-1}$$

To show this, let

$$C = \frac{\sqrt{\det \mathbf{A}}}{\sqrt{(2\pi)^n}}$$

and calculate

$$M_{\mathbf{X}}(t) = E\left[\exp\left(\sum_{i=1}^{n} t_i X_i\right)\right]$$

$$= \int_{-\infty}^{\infty} \cdots \int_{-\infty}^{\infty} C \exp\left\{-\tfrac{1}{2}[(\mathbf{x} - \boldsymbol{\mu})^{\mathrm{T}}\mathbf{A}(\mathbf{x} - \boldsymbol{\mu}) - 2\mathbf{t}^{\mathrm{T}}\mathbf{x})]\right\} dx_1 \cdots dx_n$$

which by completing the square of the exponent

$$= \int_{-\infty}^{\infty} \cdots \int_{-\infty}^{\infty} C \exp\left\{-\tfrac{1}{2}[(\mathbf{x} - \boldsymbol{\mu} - \mathbf{A}^{-1}\mathbf{t})^{\mathrm{T}}\mathbf{A}(\mathbf{x} - \boldsymbol{\mu} - \mathbf{A}^{-1}\mathbf{t})\right.$$
$$\left. - 2\boldsymbol{\mu}^{\mathrm{T}}\mathbf{t} - \mathbf{t}^{\mathrm{T}}\mathbf{A}^{-1}\mathbf{t}]\right\} dx_1 \cdots dx_n$$

$$= \exp\left(\boldsymbol{\mu}^{\mathrm{T}}\mathbf{t} + \tfrac{1}{2}\mathbf{t}^{\mathrm{T}}\mathbf{A}^{-1}\mathbf{t}\right) \int_{-\infty}^{\infty} \cdots \int_{-\infty}^{\infty} C \exp\left\{-\tfrac{1}{2}[(\mathbf{x} - \boldsymbol{\mu} - \mathbf{A}^{-1}\mathbf{t})^{\mathrm{T}}\mathbf{A}\right.$$
$$\left. \times (\mathbf{x} - \boldsymbol{\mu} - \mathbf{A}^{-1}\mathbf{t})]\right\} dx_1 \cdots dx_n$$

$$= \exp\left(\boldsymbol{\mu}^{\mathrm{T}}\mathbf{t} + \tfrac{1}{2}\mathbf{t}^{\mathrm{T}}\mathbf{A}^{-1}\mathbf{t}\right)$$

$$= \exp\left(\sum_{i=1}^{n} \mu_i t_i + \tfrac{1}{2}\sum_{i=1}^{n}\sum_{j=1}^{n} r_{ij} t_i t_j\right)$$

Using this, we find that

$$E[X_i] = \left.\frac{dM_{\mathbf{X}}(\mathbf{t})}{dt_i}\right|_{t_1 = \cdots = t_n = 0} = \mu_i$$

and

$$E[X_i X_j] = \frac{d^2 M_{\mathbf{X}}(\mathbf{t})}{dt_i \, dt_j} \Bigg|_{t_1 = \cdots = t_n = 0} = r_{ij} + \mu_i \mu_j$$

which gives

$$
\begin{aligned}
\text{cov } (X_i, X_j) &= E[(X_i - \mu_i)(X_j - \mu_j)] \\
&= E[X_i X_j] - \mu_i E[X_j] - \mu_j E[X_i] + \mu_i \mu_j \\
&= r_{ij} + \mu_i \mu_j - \mu_i \mu_j - \mu_i \mu_j + \mu_i \mu_j \\
&= r_{ij}
\end{aligned}
$$

verifying equation (0.6.3). The matrix $\mathbf{A}^{-1} = (r_{ij})$ is referred to as the **covariance matrix.** Thus the density of a family of jointly normal random variables is completely determined by its means and covariances. By considering the form of the moment generating function, we also find the important fact that jointly normal random variables X_i are independent iff the covariances $r_{ij} = 0$ for $i \neq j$.

For the two-dimensional case, suppose that X and Y are jointly normally distributed and var $X =$ var $Y = 1$. Setting $r =$ cov (X, Y), since cov $(X, X) =$ var $X = 1$, cov $(Y, Y) =$ var $Y = 1$, we have

$$\mathbf{A}^{-1} = \begin{bmatrix} 1 & r \\ r & 1 \end{bmatrix}$$

This implies that

$$\mathbf{A} = \frac{1}{1 - r^2} \begin{bmatrix} 1 & -r \\ -r & 1 \end{bmatrix}$$

so that the joint density of X and Y is

$$
f_{X,Y}(x, y) = \frac{1}{2\pi \sqrt{1 - r^2}}
$$
$$
\times \exp\left\{ -\frac{1}{2} \frac{1}{1 - r^2} [(x - \mu_X)^2 - 2r(x - \mu_X)(y - \mu_Y) + (y - \mu_Y)^2] \right\}
$$

By a change of variables we get the formula for the general bivariate normal density,

$$
f_{X,Y}(x,y) = \frac{1}{2\pi \sigma_X \sigma_Y \sqrt{1 - \rho^2}} \tag{0.6.5}
$$
$$
\times \exp\left\{ -\frac{1}{2} \frac{1}{1 - \rho^2} \left[\frac{(x - \mu_X)^2}{\sigma_X^2} - 2\rho \frac{(x - \mu_X)(y - \mu_Y)}{\sigma_X \sigma_Y} + \frac{(y - \mu_Y)^2}{\sigma_Y^2} \right] \right\}
$$

where $\rho =$ cov $(X, Y)/\sigma_X \sigma_Y$ is the correlation coefficient of X and Y.

For joint normal random variables the conditional expectation has a nice inter-pretation. If the X_i's have mean zero, the conditional expectation

$$E[X_n | X_{n-1} = x_{n-1}, \ldots, X_1 = x_1]$$

$$= \int_{-\infty}^{\infty} x_n f(x_n | X_{n-1} = x_{n-1}, \ldots, X_1 = x_1) \, dx_n$$

can be characterized as the unique linear function

$$E[X_n | X_{n-1} = x_{n-1}, \ldots, X_1 = x_1] = b_{n-1} x_{n-1} + \cdots + b_1 x_1$$

such that the corresponding random variable

$$X_n - E[X_n | X_{n-1}, \ldots, X_1] = X_n - b_{n-1} X_{n-1} - \cdots - b_1 X_1$$

is independent of $X_{n-1}, \ldots, X_1$. To show this, first note that the conditional density of X_n given $X_{n-1}, \ldots, X_1$ is

$$f(x_n | X_{n-1} = x_{n-1}, \ldots, X_1 = x) = \frac{f_X(x_1, \ldots, x_n)}{\int_{-\infty}^{\infty} f_X(x_1, \ldots, x_n) \, dx_n}$$

or

$$= \frac{\exp\left(-\dfrac{1}{2} a_{nn} x_n^2 - \sum_{i=1}^{n-1} a_{in} x_i x_n\right)}{\int_{-\infty}^{\infty} \exp\left(-\dfrac{1}{2} a_{nn} x_n^2 - \sum_{i=1}^{n-1} a_{in} x_i x_n\right) dx_n}$$

which by completing the square

$$= \frac{\exp\left\{-\dfrac{1}{2} a_{nn}\left[x_n + \sum_{i=1}^{n-1} (a_{in}/a_{nn}) x_i\right]^2\right\}}{\int_{-\infty}^{\infty} \exp\left\{-\dfrac{1}{2} a_{nn}\left[x_n + \sum_{i=1}^{n-1} (a_{in}/a_{nn}) x_i\right]^2\right\} dx_n}$$

and noting that the denominator does not depend on $x_1, \ldots, x_n$,

$$= K \exp\left\{-\dfrac{1}{2} a_{nn}\left[x_n + \sum_{i=1}^{n-1} (a_{in}/a_{nn}) x_i\right]^2\right\}$$

where K is a normalizing constant depending on (a_{ij}) but independent of $x_1, \ldots,$ x_n. As a function of x_n this is a normal density, with mean

$$-\sum_{i=1}^{n-1} \frac{a_{in}}{a_{nn}} x_i$$

so

$$E[X_n | X_{n-1} = x_{n-1}, \ldots, X_1 = x_1] = -\sum_{i=1}^{n-1} \frac{a_{in}}{a_{nn}} x_i$$

Therefore, the conditional expectation is a linear combination as above, with

$$b_k = -\frac{a_{kn}}{a_{nn}}$$

Now consider the random variables

$$Y_n = X_n - \sum_{j=1}^{n-1} b_j X_j$$

and

$$Y_j = X_j, \qquad j = 1, \ldots, n-1$$

Letting $\mathbf{Z} = (X_1, \ldots, X_{n-1})$, our calculations above show that

$$f_{\mathbf{X}}(x_1, \ldots, x_n) = K \exp\left[-\tfrac{1}{2} a_{nn} \left(x_n - \sum_{j=1}^{n-1} b_j x_j \right)^2 \right] f_{\mathbf{Z}}(x_1, \ldots, x_{n-1})$$

so the density for $\mathbf{Y}$ factors

$$f_{\mathbf{Y}}(y_1, \ldots, y_n) = f_{Y_n}(y_n) f_{Y_1, \ldots, Y_{n-1}}(y_1, \ldots, y_{n-1})$$

Hence

$$Y_n = X_n - B_{n-1} X_{n-1} - \cdots - b_1 X_1$$

is independent of every

$$Y_j = X_j, \qquad j = 1, \ldots, n-1$$

as required. In fact, we can see that Y_n has a normal distribution.

To show uniqueness, let $b_{n-1}x_{n-1} + \cdots + b_1 x_1$ be a linear combination such that $X_n - b_{n-1}X_{n-1} - \cdots - b_1 X_1$ is independent of $X_1, \ldots, X_{n-1}$. This implies that for $i = 1, \ldots, n - 1$,

$$E[X_i(X_n - b_{n-1}X_{n-1} - \cdots - b_1 X_1)] = E[X_i]E[X_n - b_{n-1}X_{n-1} - \cdots - b_1 X_1]$$
$$= 0$$

or, in terms of the covariance matrix $(r_{ij}) = \mathbf{R} = \mathbf{A}^{-1}$,

$$r_{in} - b_{n-1}r_{in-1} - \cdots - b_1 r_{i1} = 0, \qquad i = 1, \ldots, n - 1$$

Since $\mathbf{R}$ has independent rows, the b_k's are unique.

We shall see later that arbitrary linear combinations of jointly distributed normal random variables are normal. Thus it is possible to define a vector space of mean zero normal random variables, consisting of all linear combinations of mean zero normal random variables. If this vector space is given the inner product defined by

$$<X, Y> = E[XY]$$

then the notions of independence for random variables and orthogonality for vectors are the same. Further, by what we have just seen, the conditional expectation

$$E[X|Y_1, \ldots, Y_n]$$

corresponds to the orthogonal projection onto the subspace spanned by $\{Y_1, \ldots, Y_n\}$, since it is exactly what must be subtracted from X to make it orthogonal to every Y_i.

0.7   SUMS OF INDEPENDENT RANDOM VARIABLES

Sum of two uniform random variables

We first consider the sum of two independent uniform random variables X and Y with common density function

$$f(x) = \begin{cases} 1, & 0 \le x \le 1 \\ 0, & \text{else} \end{cases}$$

Letting $Z = X + Y$, we calculate

$$P[Z \le z] = \int_0^z \int_0^{z-y} f(x)f(y) \, dx \, dy$$

$$= \int_0^z \int_0^{z-y} I_{[0,1]}^2 \, dx \, dy$$

$$= \begin{cases} -\frac{1}{2}z^2 + 2z - 1, & z \ge 1 \\ \frac{1}{2}z^2, & z < 1 \end{cases}$$

Differentiating this, we find the density function for Z to be

$$f_Z(z) = \begin{cases} 2 - z, & z \ge 1 \\ z, & z < 1 \end{cases}$$

Sum of n exponential random variables

Let us calculate the distribution of a slightly more complicated example. Suppose that $X_1, \ldots, X_n$ are independent identically distributed exponential random variables with density

$$f_X(x) = \begin{cases} e^{-x}, & x \ge 0 \\ 0, & x < 0 \end{cases}$$

Let $Z = X_1 + \cdots + X_n$, so that

$$P[Z \le z] = \int_0^z \int_0^{z-x_n} \cdots \int_0^{z-x_2-\cdots-x_n} \exp\left(-\sum_{i=1}^n x_i\right) dx_1 \cdots dx_n$$

$$= \int_0^z \int_0^{z-x_n} \cdots \int_0^{z-x_3-\cdots-x_n} \exp\left(-\sum_{i=2}^n x_i\right) - e^{-z} \, dx_2 \cdots dx_n$$

Noting that one can show by induction

$$\int_0^z \int_0^{z-x_n} \cdots \int_0^{z-x_2-\cdots-x_n} dx_1 \cdots dx_n = \frac{z^n}{n!} \tag{0.7.1}$$

we obtain

$$P[Z \le z] = \frac{-e^{-z}z^{n-1}}{(n-1)!} + \int_0^z \int_0^{z-x_n} \cdots \int_0^{z-x_3-\cdots-x_n} \exp\left(-\sum_{i=2}^n X_i\right) dx_2 \cdots dx_n$$

This formula allows us to find the distribution for $n = k + 1$ if we know it for $n = k$. Since for $n = 1$,

$$P[Z \le z] = -e^{-z} + 1$$

then for $n = 2$,

$$P[Z \leq z] = -e^{-z}z - e^{-z} + 1$$

and for $n = 3$,

$$P[Z \leq z] = \frac{e^{-z}z^2}{2} + -e^{-z}z - e^{-z} + 1$$

and, in general, for $n = k$,

$$P[Z \leq z] = 1 - e^{-z}\left[1 + z + \frac{z^2}{2!} + \frac{z^3}{3!} + \cdots + \frac{z^{k-1}}{(k-1)!}\right]$$

Differentiating this gives the density

$$f_Z(z) = \frac{z^{k-1}e^{-z}}{(k-1)!} \tag{0.7.2}$$

The gamma densities

The density (0.7.2) is one of the family of gamma densities, with parameters (t, λ), defined by the formula

$$f_X(x) = \frac{\lambda e^{-\lambda x}(\lambda x)^{t-1}}{\Gamma(t)}, \qquad 0 < x < \infty \tag{0.7.3}$$

where Γ is the gamma function defined by

$$\Gamma(t) = \int_0^\infty e^{-y}y^{t-1}\,dy \tag{0.7.4}$$

We note that $\Gamma(n) = (n-1)!$ if n is a positive integer.

The convolution formula for the sum of two independent random variables

Suppose that X and Y are two independent random variables, with densities f_X and f_Y, respectively. If we let $Z = X + Y$, then since the joint density of X and Y is $f_X f_Y$,

$$P[Z \leq z] = \int_{-\infty}^\infty \int_{-\infty}^{z-x} f_X(x)f_Y(y)\,dy\,dx$$

The density of Z is

$$\frac{d}{dz} P[Z < z] = \frac{d}{dz}\left[\int_{-\infty}^{\infty} \int_{-\infty}^{z-x} f_X(x)f_Y(y)\, dy\, dx\right]$$

$$= \int_{-\infty}^{\infty} f_X(x)f_Y(z - x)\, dx$$

or

$$f_Z(z) = f_X * f_Y(z) \tag{0.7.5}$$

We shall use this formula to show that the sum of two independent gamma random variables is a gamma random variable.

Sum of two independent gamma random variables

Suppose that X and Y are independent gamma random variables with parameters (s, λ) and (t, λ), respectively. By (0.7.5) the density function for $Z = X + Y$ is

$$f_Z(z) = \frac{1}{\Gamma(t)\Gamma(s)} \int_0^z \lambda e^{-\lambda x}(\lambda x)^{t-1}\lambda e^{-\lambda(z-x)}[\lambda(z - x)]^{s-1}\, dx$$

for some constant C_1,

$$= C_1 e^{-\lambda z} \int_0^z x^{t-1}(z - x)^{s-1}\, dx$$

or for some constant C_2,

$$= C_2 e^{-\lambda z} z^{t+s-1}$$

Since this is a probability density, the normalizing constant C_2, when compared with (0.7.3), must of course be $1/\Gamma(t + s)$. Thus Z is gamma with parameters $(t + s, \lambda)$. This also shows that the sum of n independent gamma random variables with parameters (t_k, λ) is a gamma random variable with parameters $(\Sigma t_k, \lambda)$.

Use of the moment generating function

If X and Y are independent random variables, then e^{tX} and e^{tY} will be independent, and we can conclude that in this case,

$$M_{X+Y}(t) = M_X(t)M_Y(t)$$

More generally, if $X_1, \ldots, X_n$ is a family of independent random variables, and $Z = X_1 + \cdots + X_n$,

$$M_Z(t) = \prod_{i=1}^{n} M_{X_i}(t) \tag{0.7.6}$$

This can be exploited in many cases to determine the distribution of sums of independent random variables.

Linear combinations of normal random variables

If X and Y are independent normally distributed, we can show that any linear combination $aX + bY$ is a normal random variable. Since aX and bY are independent, by (0.7.6)

$$
\begin{aligned}
M_{aX+bY} &= M_{aX} M_{bY} \\
&= \exp\left(a\mu_X t + \tfrac{1}{2}a^2\sigma_X^2 t^2\right) \exp\left(b\mu_Y t + \tfrac{1}{2}b^2\sigma_Y^2 t^2\right) \\
&= \exp\left[(a\mu_X + b\mu_Y)t + \tfrac{1}{2}(a^2\sigma_X^2 + b^2\sigma_Y^2)t^2\right]
\end{aligned}
$$

This is the moment generating function of a normal random variable of mean $a\mu_X + b\mu_Y$ and variance $a^2 \operatorname{var} X + b^2 \operatorname{var} Y$; therefore, $aX + bY$ is normal with this mean and variance.

The chi-squared distribution

If $X_1, \ldots, X_n$ are independent normal random variables of mean zero and variance 1,

$$Y = X_1^2 + \cdots + X_n^2 \tag{0.7.7}$$

is called a chi-squared random variable with n degrees of freedom. This random variable is important in statistics, and we now discuss it in more detail.

Consider first the case $n = 1$, and find the distribution of $Y = X_1^2$:

$$
\begin{aligned}
F_Y(y) &= P[X_1^2 \le y] = P[-\sqrt{y} \le X_1 \le \sqrt{y}] \\
&= 2P[0 \le X_1 \le \sqrt{y}] \\
&= 2 \int_0^{\sqrt{y}} \frac{1}{\sqrt{2\pi}} \exp\left(-\tfrac{1}{2}x^2\right) dx
\end{aligned}
$$

From this we can find the density of Y:

$$f_Y(y) = \frac{dF_Y(y)}{dy}$$

$$= 2 \frac{1}{\sqrt{2\pi}} e^{-(1/2)y} \frac{d\sqrt{y}}{dy}$$

$$= \frac{\frac{1}{2} e^{-(1/2)y} (\frac{1}{2}y)^{(1/2)-1}}{\sqrt{\pi}}$$

which says that Y has a gamma density with parameters $(\frac{1}{2}, \frac{1}{2})$. We can also find its moment generating function:

$$M_Y(t) = E[\exp(tX^2)]$$

$$= \int_{-\infty}^{\infty} \exp(tx^2) \frac{1}{\sqrt{2\pi}} \exp(-\tfrac{1}{2}x^2) \, dx$$

$$= \int_{-\infty}^{\infty} \frac{1}{\sqrt{2\pi}} \exp[-\tfrac{1}{2}x^2(1 - 2t)] \, dx$$

$$= \frac{1}{\sqrt{1 - 2t}} \int_{-\infty}^{\infty} \frac{1}{\sqrt{2\pi}} \exp(-\tfrac{1}{2}w^2) \, dw$$

$$= \frac{1}{\sqrt{1 - 2t}}$$

From these we can conclude that a chi-squared random variable with n degrees of freedom has a gamma density with parameters $(n/2, \frac{1}{2})$

$$f_Y(y) = \frac{\frac{1}{2} e^{-(1/2)y} (\frac{1}{2}y)^{(n/2)-1}}{\Gamma(n/2)} \tag{0.7.8}$$

and moment generating function

$$M_Y(t) = \left(\frac{1}{\sqrt{1 - 2t}}\right)^n \tag{0.7.9}$$

0.8 CONVERGENCE OF RANDOM VARIABLES

Convergence in law

A family of random variables X_n is said to converge to a random variable Y in law if

$$\lim_{n \to \infty} F_{X_n}(t) = F_Y(t) \tag{0.8.1}$$

for any t where F_Y is continuous. The distribution $F_Y(t)$ is then called the **limiting distribution** of the family X_n.

Central limit theorem

The central limit theorem says that the sum of many independent identically distributed random variables, once suitably scaled, is always approximately normally distributed. We state it below and give a proof in Chapter 1.

Let X_1, X_2, . . . be an independent family of identically distributed random variables with mean μ and variance σ^2. If

$$S_n = X_1 + \cdots X_n$$

and

$$W_n = \frac{S_n - n\mu}{\sigma\sqrt{n}} \tag{0.8.2}$$

then W_n converges in law to a normal random variable of mean zero and variance 1, that is,

$$\lim_{n\to\infty} P[W_n \le t] = \int_{-\infty}^{t} \frac{1}{\sqrt{2\pi}} e^{-(1/2)x^2} \, dx \tag{0.8.3}$$

As an example of this convergence, below we find the limiting distribution for sums of independent identical exponential random variables. After this we find the limiting distribution for binomial random variables, which are sums of independent identical Bernoulli random variables. The central limit theorem can be generalized to sums of independent random variables which are not identically distributed, provided that they are about the same "size." Since many commonly observed random quantities are the result of adding many small independent quantities together, this explains the importance of the normal distribution in many practical situations.

Let X_1, . . . , X_n be independent identically distributed exponential random variables with density

$$f_X(x) = \begin{cases} e^{-x}, & x \ge 0 \\ 0, & x < 0 \end{cases}$$

and let

$$W = \frac{\displaystyle\sum_{i=1}^{n} X_i - n}{\sqrt{n}}$$

The distribution function for W is

$$F_W(w) = P\left[\frac{\sum_{i=1}^{n} X_i - n}{\sqrt{n}} \le w\right]$$

$$= P\left[\sum_{i=1}^{n} X_i \le w\sqrt{n} + n\right]$$

or, according to (0.7.2),

$$= \int_0^{w\sqrt{n}+n} \frac{z^{n-1}e^{-z}}{(n-1)!}\, dz$$

Differentiating this gives the density function for W:

$$f_W(w) = \frac{(w\sqrt{n} + n)^{n-1} \exp\left[-(w\sqrt{n} + n)\right]}{(n-1)!} \sqrt{n}$$

and using Stirling's formula, this is approximately

$$\frac{(w\sqrt{n} + n)^{n-1} \exp\left[-(w\sqrt{n} + n)\right]}{n^{n-1} \exp(-n) \sqrt{2\pi n}} \sqrt{n}$$

$$= \frac{1}{\sqrt{2\pi}}\left(\frac{w}{\sqrt{n}} + 1\right) \exp\left[-w\sqrt{n} + n \ln\left(\frac{w}{\sqrt{n}} + 1\right)\right]$$

Thus one can check by means of the Taylor's series expansion

$$\ln(1 + x) = x - \frac{x^2}{2} + \frac{x^3}{3} + \frac{x^4}{4} + \cdots$$

that in the limit f_W becomes a normal density

$$\lim_{n \to \infty} f_W(w) = \frac{1}{\sqrt{2\pi}} \exp\left(-\tfrac{1}{2}w^2\right)$$

Now we find the limiting distribution for binomial random variables. This result is sometimes called the DeMoivre–Laplace limit theorem.

Consider the binomial random variable B_n,

$$P[B_n = k] = \binom{n}{k}\left(\tfrac{1}{2}\right)^n$$

which is the sum of n independent identical Bernoulli random variables, satisfying

$$P[X_i = 1] = P[X_i = 0] = \tfrac{1}{2}$$

and let

$$W = \frac{B_n - n/2}{\sqrt{n}/2}$$

The distribution function of W is

$$P[W \le w] = P[2B_n - n \le w\sqrt{n}]$$
$$= P[B_n \le \tfrac{1}{2}(w\sqrt{n} + n)]$$

which is a sum over the nonnegative integers $k \le (1/2)(w\sqrt{n} + n)$

$$= \sum_{k=0}^{(1/2)(w\sqrt{n}+n)} (\tfrac{1}{2})^n \binom{n}{k}$$

By Stirling's formula, this is approximately

$$= \sum_{k=0}^{(1/2)(w\sqrt{n}+n)} \frac{1}{\sqrt{2\pi}} \frac{2}{\sqrt{n}} \exp\left[-\frac{1}{2}\left(\frac{2k-n}{\sqrt{n}}\right)^2 \right]$$

which we note is a Riemann sum approximating the integral

$$\int_{-\sqrt{n}}^{w} \frac{1}{\sqrt{2\pi}} \exp\left(-\tfrac{1}{2}x^2\right) dx$$

and as $n \to \infty$ this becomes

$$\int_{-\infty}^{w} \frac{1}{\sqrt{2\pi}} \exp\left(-\tfrac{1}{2} x^2\right) dx$$

Thus for large n, B_n, once suitably normalized, is approximately normally distributed.

Another way of verifying convergence in law is to use the following fact about the moment generating function. If there exists a sequence of random variables X_n and a random variable Y such that their moment generating functions exist and satisfy

$$\lim_{n\to\infty} M_{X_n}(t) = M_Y(t) \tag{0.8.4}$$

for each t, then X_n converges to Y in law.

We use this to show that if N_λ is a Poisson random variable of parameter λ, then

$$W = \frac{N_\lambda - \lambda}{\sqrt{\lambda}}$$

converges in law, as $\lambda \to \infty$, to a normal random variable of mean zero and variance 1. This cannot be concluded directly from the central limit theorem, but since a Poisson random variable is approximately binomial one might expect this result. The moment generating function for W is

$$M_W(t) = E[\exp{(tW)}]$$

$$= \sum_{k=0}^{n} \exp{\left(t\frac{k - \lambda}{\sqrt{\lambda}}\right)}\frac{e^{-\lambda}\lambda^k}{k!}$$

$$= \sum_{k=0}^{n} e^{-\lambda}e^{-t\sqrt{\lambda}}\frac{(\lambda e^{t/\sqrt{\lambda}})^k}{k!}$$

$$= e^{-\lambda}e^{-t\sqrt{\lambda}}\exp{(\lambda e^{t/\sqrt{\lambda}})}$$

$$= \exp{(-\lambda - t\sqrt{\lambda} + \lambda e^{t/\sqrt{\lambda}})}$$

Writing

$$-\lambda - t\sqrt{\lambda} + \lambda e^{t/\sqrt{\lambda}} = -\lambda - t\sqrt{\lambda} + \lambda\left[1 + \frac{t}{\sqrt{\lambda}} + \frac{(t/\sqrt{\lambda})^2}{2!} + \cdots\right]$$

$$= \frac{t^2}{2!} + \frac{t^3}{3!\sqrt{\lambda}} + \cdots$$

shows that as $\lambda \to \infty$,

$$\lim_{\lambda \to \infty} M_W(t) = \exp{(\tfrac{1}{2}t^2)}$$

which is the moment generating function of a normal random variable of mean zero and variance 1.

Convergence in probability

A family of random variables X_n is said to converge to a random variable Y in probability if

$$\lim_{n \to \infty} P[|X_n - Y| > \epsilon] = 0 \qquad (0.8.5)$$

for any $\epsilon > 0$. This is a stronger kind of convergence than convergence in law, since the values of the random variables are forced to be close together, not just their distributions. It can be shown that convergence in probability implies convergence in law.

Weak law of large numbers

The average of a large number of independent identically distributed random variables is very likely to be near the mean of an individual random variable. This is made precise by the following, which is referred to as the weak law of large numbers.

Let $X_1, X_2, \ldots$ be an independent family of identically distributed random variables with mean μ and variance σ^2. Then

$$Y_n = \frac{X_1 + \cdots + X_n}{n} \tag{0.8.6}$$

converges to μ in probability, that is,

$$\lim_{n \to \infty} P[|Y_n - \mu| > \epsilon] = 0 \qquad \text{for any } \epsilon > 0 \tag{0.8.7}$$

The proof of this turns out to be an easy application of Chebyshev's inequality. Recall that Chebyshev's inequality says that

$$P[|X - E[X]| \geq \epsilon] \leq \frac{\text{var } X}{\epsilon^2}$$

Setting X to be the average

$$X = \frac{1}{n} \sum_{k=1}^{n} X_k$$

this becomes

$$P\left[\left|\frac{1}{n} \sum_{k=1}^{n} X_k - \mu\right| \geq \epsilon\right] \leq \frac{\sigma^2}{n\epsilon^2}$$

and the result follows.

As a surprizing and elegant application of the weak law of large numbers, we give a proof, due to S. Bernstein, of the Weierstrass approximation theorem which says that any continuous function f on $[0, 1]$ can be approximated uniformly by polynomials. Consider the "Bernstein" polynomials associated with a function f defined on $[0, 1]$:

$$P_n(x) = \sum_{k=0}^{n} f\left(\frac{k}{n}\right) \binom{n}{k} x^k (1 - x)^{n-k}$$

This is, in fact, the expected value $E[f(Y/n)]$, where Y is a binomial random variable with parameter x. Since f is continuous on $[0, 1]$, it is necessarily uniformly continuous on $[0, 1]$ as well. Since Y is the sum of n identically distributed independent Bernoulli random variables, the proof of the weak law of large numbers says that $P[|Y/n - x| > \epsilon] \to 0$ as $n \to \infty$, uniformly in x. Combining these, it is easy to show that $P_n \to f$, uniformly in $[0, 1]$.

Almost sure convergence

Almost sure convergence is an even stronger type of convergence than convergence in probability, stronger meaning that it implies convergence in probability, which in turn implies convergence in law. A sequence of random variables X_n is said to converge to Y almost surely if for each $\epsilon > 0$,

$$\lim_{N \to \infty} P[|X_n - Y| < \epsilon \quad \text{for all } n \geq N] = 1$$

Almost sure convergence is abbreviated $X_n \to Y$ a.s.

The Borel–Cantelli lemmas

An important tool for showing almost sure convergence is use of the Borel–Cantelli lemmas. Before stating these, we give a result from the theory of infinite products.
Suppose that $0 \leq x_n < 1$; then

$$\prod_{k=1}^{\infty} (1 - x_k) = 0$$

precisely when

$$\sum_{k=1}^{\infty} x_k = \infty$$

This is easy to see from the comparisons

$$1 - x \leq e^{-x}$$

for all x, and

$$1 - x \geq e^{-2x}$$

for $0 \leq x < \frac{1}{2}$.
A second result needed is the **continuity of probability,** by which we mean that if E_k is an increasing or decreasing sequence of events,

$$E_1 \subset \cdots \subset E_k \subset \cdots$$

or

$$E_1 \supset \cdots \supset E_k \supset \cdots$$

then

$$P\left[\lim_{k \to \infty} E_k\right] = \lim_{k \to \infty} P[E_k]$$

The proof of this is left as an exercise.

Given a sequence of events $E_1, E_2, \ldots$, the event that they happen infinitely often is defined by

$$\bigcap_{n=1}^{\infty} \bigcup_{k=n}^{\infty} E_k$$

As an application of our result on infinite products, we can show that if $E_1, \ldots, E_n, \ldots$ are independent events, then the probability that they happen infinitely often is 1 if

$$\sum_{k=1}^{\infty} P[E_k] = \infty$$

This is the **first Borel–Cantelli lemma.** To do this, consider the probability that none of the E_k's for $k \geq n$ happen:

$$P\left[\bigcap_{k=n}^{\infty} E_k^c\right] = \prod_{k=n}^{\infty} P[E_k^c] = \prod_{k=n}^{\infty} (1 - P[E_k])$$

If the sum is infinite, the product is zero, indicating that

$$P\left[\bigcup_{k=n}^{\infty} E_k\right] = 1$$

which implies the result, by continuity of probability.

Even without independence, finiteness of the sum

$$\sum_{k=1}^{\infty} P[E_k] < \infty$$

implies that the events cannot happen infinitely often. This is the **second Borel–Cantelli lemma.** Finiteness implies that the partial sums

$$\sum_{k=n}^{\infty} P[E_k]$$

approach zero as $n \to \infty$. Since this is a bound on the probability of at least one E_k happening, for $k \geq n$,

$$P\left[\bigcup_{k=n}^{\infty} E_k\right] \geq \sum_{k=n}^{\infty} P[E_k]$$

the result is clear.

Suppose we wish to prove that X_n converges to Y a.s. By letting E_k be the event

$$E_k = \{|X_k - Y| > \epsilon\}$$

if

$$\sum_{k=1}^{\infty} P[E_k] < \infty$$

for any ϵ, then by the second Borel–Cantelli lemma, as $n \to \infty$

$$P[|X_k - Y| > \epsilon \text{ for some } k \geq n] \to 0$$

which says that the sequence converges almost surely to Y.

The strong law of large numbers

Suppose that $X_1, \ldots, X_n, \ldots$ is a sequence of identically distributed random variables. The strong law of large numbers says that there exists a constant μ such that

$$\frac{1}{n} \sum_{k=1}^{n} X_k \to \mu \qquad \text{a.s.}$$

precisely when $E[|X_i|] < \infty$, and in this case $\mu = E[X_i]$. Further if the $X_k \geq 0$ and $E[X_i] = \infty$, then the sum approaches infinity.

We shall show this in the case that X_i has a finite second moment. The proof is based on the Doob-Kolmogorov inequality for martingales, which is a generalization of Chebyshev's inequality and is given below. First we must introduce the useful and important concept of a martingale.

A sequence of random variables Y_n is called a **martingale** if

$$E[Y_{n+1}|Y_n, \ldots, Y_1] = Y_n$$

which implies that

$$E[Y_{n+k}|Y_n, \ldots, Y_1] = Y_n$$

One important example of a martingale is obtained by considering the partial sums of independent identically distributed mean zero random variables X_i:

$$Y_n = \sum_{k=1}^{n} X_k$$

but there are many other examples. Martingales satisfy the **Doob–Kolmogorov inequality,** which says that if Y_n is a martingale and $E[Y_n^2] < \infty$, then

$$P\left[\max_{1 \le k \le n} |Y_k| \ge \epsilon\right] \le \frac{1}{\epsilon^2} E[Y_n^2]$$

A quick sketch of the proof is as follows. Define events E and E_i, $i = 1, \ldots, n$ by

$$E_i = \{|Y_k| \ge \epsilon \text{ for the first time when } k = i\}$$

and

$$E = \bigcup_{i=1}^{n} E_i$$

Now

$$E[Y_n^2] \ge E[Y_n^2 \, 1_E]$$

$$= \sum_{i=1}^{n} E[Y_n^2 \, 1_{E_i}]$$

$$= \sum_{i=1}^{n} E[(Y_n - Y_i + Y_i)^2 1_{E_i}]$$

$$= \sum_{i=1}^{n} E[(Y_n - Y_i)^2 1_{E_i}] + 2E[(Y_n - Y_i)Y_i 1_{E_i}] + E[Y_i^2 1_{E_i}]$$

Since

$$E[(Y_n - Y_i)Y_i 1_{E_i}] = E[E(Y_n - Y_i)Y_i 1_{E_i} | Y_i, \ldots, Y_1]$$

$$= E[Y_i 1_{E_i} E[Y_n - Y_i | Y_i, \ldots, Y_1]]$$

$$= 0$$

by the martingale condition, we conclude that

$$E[Y_n^2] \ge \sum_{i=1}^{n} E[Y_i^2 1_{E_i}] \ge \epsilon^2 \sum_{i=1}^{n} P[E_i] = \epsilon^2 P\left[\max_{1 \le k \le n} |Y_k| \ge \epsilon\right]$$

which completes the proof.

We can use this inequality to give a proof of the strong law of large numbers in the case that X_i has finite second moment. Assume for convenience that X_i has mean zero. Let S_n be the partial sum

$$S_n = \sum_{k=1}^{n} X_k$$

We will show that for each $\epsilon > 0$

$$P\left[\left|\frac{S_n}{n}\right| \geq \epsilon \text{ infinitely often}\right] = 0$$

which says that

$$P\left[\lim_{n \to \infty} \frac{S_n}{n} = 0\right] = 1$$

Let

$$D(n) = \min (d|d = 2^k, 2^k \geq n, k = 0, 1, 2, \ldots)$$

be the closest power of 2 greater than or equal to n. Since

$$\left|\frac{S_n}{n}\right| \leq 2 \left|\frac{S_n}{D(n)}\right|$$

it is sufficient to show

$$P\left[\left|\frac{S_n}{D(n)}\right| \geq \frac{\epsilon}{2} \text{ infinitely often}\right] = 0 \qquad (0.8.8)$$

To do this we use the second Borel–Cantelli lemma. For each m, consider the martingale

$$Y_n^{(m)} = \frac{1}{2^m} \sum_{k=1}^{n} X_k$$

The Doob–Kolmogorov inequality says that

$$P\left[\max_{1 \leq k \leq 2^m} |Y_k^{(m)}| \geq \frac{\epsilon}{2}\right] \leq \frac{4}{\epsilon^2} \frac{\sigma^2}{2^m}$$

and summing over m we see that this is finite, so

$$P\left[\max_{1 \le k \le 2^m} |Y_k^{(m)}| \ge \frac{\epsilon}{2} \text{ for infinitely many } m\right] = 0$$

This implies (0.8.8).

Another important property of martingales is the **martingale convergence theorem,** which says that if

$$E[Y_n^2] < M < \infty$$

then the martingale converges almost surely to a random variable Y. This is also a consequence of the Doob–Kolmogorov inequality, but we omit a proof.

Mean square convergence

A family of random variables X_n is said to converge to Y in mean square if $E[Y^2] < \infty$ and

$$\lim_{n \to \infty} E[(X_n - Y)^2] = 0$$

This is usually abbreviated l.i.m. $X_n = Y$.

Mean square convergence implies convergence in probability, and thus convergence in law. There are almost sure convergent sequences that are not mean square convergent, and vice versa.

Limits and expectation

Expectation is a form of integration, and therefore part of measure theory. A mathematically rigorous discussion of probability theory would include measure theory as the unifying setting for the concepts involved. However, the understanding of probability theory lies in understanding the examples, which for the most part is best done without reference to the general theory.

In some instances we need to know conditions under which limit and expectation may be interchanged. The following are just the usual theorems of integration theory, stated in terms of expectation. They can be shown by appealing to the general theory. The proofs will not be presented here.

The monotone convergence theorem:

> If $X_n \ge 0$ and X_n increase a.s. to a random variable X, then
> $$\lim_{n \to \infty} E[X_n] = E[X]$$

The bounded convergence theorem:

If $\lim_{n \to \infty} X_n = X$ a.s. and $|X_n| < M =$ a constant $< \infty$, then
$$\lim_{n \to \infty} E[X_n] = E[X]$$

The dominated convergence theorem:

If $\lim_{n \to \infty} X_n = X$ a.s. and $|X_n| < Y$ where $E[Y] < \infty$, then
$$\lim_{n \to \infty} E[X_n] = E[X]$$

Fatou's lemma:

For any sequence of random variables X_n, $E[\varliminf X_n] \le \varliminf E[X_n]$, where $\varliminf$ of a sequence x_n is defined by
$$\varliminf x_n = \lim_{N \to \infty} \inf\{x_n | n \ge N\}$$

PROBLEMS

0.1. Show Boole's inequalities:

(a) $P\left[\bigcup_{i=1}^{n} A_i\right] \le \sum_{i=1}^{n} P[A_i]$

(b) $P\left[\bigcap_{i=1}^{n} A_i\right] \ge 1 - \sum_{i=1}^{n} P[A_i]$

0.2. Using the relations (0.1.13), prove by induction DeMorgan's formulas:

(a) $\left(\bigcup_{i=1}^{n} A_i\right)^c = \bigcap_{i=1}^{n} A_i^c$

(b) $\left(\bigcap_{i=1}^{n} A_i\right)^c = \bigcup_{i=1}^{n} A_i^c$

Use these to show the relation

(c) $1_A = \sum_{i} 1_{A_i} - \sum_{i>j} 1_{A_i \cap A_j} + \cdots + (-1)^{n+1} 1_{A_1 \cap \cdots \cap A_n}$

where $a = \bigcup_{i=1}^{n} A_i$ and conclude that

(d) $P\left[\bigcup_{i=1}^{n} A_i\right] = \sum_{i} P[A_i] - \sum_{i>j} P[A_i \cap A_j] + \cdots$
$$+ (-1)^{n+1} P[A_1 \cap \cdots \cap A_n]$$

0.3. If X is a random variable, find the distribution functions of max $(0, X)$, min $(0, X)$, max $(0, X) +$ min $(0, X)$ in terms of the distribution function of X.

0.4. If X and Y are independent, show that

(a) max (X, Y) has distribution function $F_X F_Y$.

(b) min (X, Y) has distribution function $1 - (1 - F_X)(1 - F_Y)$.

0.5. If X and Y are independent random variables prove, for any value of t, that e^{tX} and e^{tY} are independent.

0.6. If X_k is a sequence of Bernoulli trials, where $P[X_k = 1] = p$ and

$$Y = \min (k | X_k = 1)$$

find the probability mass function for Y.

0.7. If X is uniformly distributed on $[0, 1]$,

(a) find the distribution function of X^2.

(b) differentiate this to find the density function of X^2.

0.8. If X, Y, and Z are independent and uniform on $[0, 1]$, find the density of $X + Y + Z$.

0.9. Suppose that $f_X(x)$ is the density function of a real random variable X, $g(x)$ is a 1-to-1 piecewise differentiable function, and $Y = g(X)$. Set $h(x) = g^{-1}(x)$. By the change of variables, $x = h(y)$, $dx = h'(y) \, dy$, we can conclude that

$$P[Y \in B] = P[X \in h(B)]$$

$$= \int_{h(B)} f_X(x) \, dx = \int_B f_X(h(y)) \, |h'(y)| \, dy$$

so that

$$f_Y(y) = f_X(h(y)) \, |h'(y)|$$

(a) Use this to show that if X is continuous, $F_X(X)$ is uniform on $[0, 1]$.

(b) Show that $-\log F_X(X)$ is exponential.

(c) If X is uniformly distributed between $-\pi$ and π, show that $\cot X$ has a Cauchy distribution.

0.10. Suppose that $\mathbf{X}$ is a vector-valued random variable with density $f_\mathbf{X}(\mathbf{x})$, $\mathbf{g} = (g_1, \ldots, g_n)$ is a 1-to-1 piecewise differentiable function and $\mathbf{Y} = \mathbf{g}(\mathbf{X})$. As in Problem 0.9, by setting $\mathbf{h} = \mathbf{g}^{-1}$ we have

$$P[\mathbf{Y} \in \mathbf{B}] = P[\mathbf{X} \in h(\mathbf{B})]$$

$$= \int_{h(B)} f_\mathbf{X}(\mathbf{x}) \, dx_1 \cdots dx_n = \int_\mathbf{B} f_\mathbf{X}(\mathbf{h}(\mathbf{y})) |\det \mathbf{h}'| \, dy_1 \cdots dy_n$$

where $\mathbf{h}'$ is the matrix of partial derivatives

$$h'_{ij} = \frac{\partial h_i}{\partial x_j}$$

so that

$$f_{\mathbf{Y}}(y_1, \ldots, y_n) = f_{\mathbf{X}}(h(\mathbf{y})) \, |\det \mathbf{h}'|$$

(a) Use this to show that if $\mathbf{h}$ is an orthogonal linear transformation, $f_{\mathbf{Y}}(\mathbf{y}) = f_{\mathbf{X}}(\mathbf{h}(\mathbf{y}))$.

(b) If X and Y are independent, with densities f_X and f_Y, respectively, find the joint density for $X - Y$ and $X + Y$.

(c) If X and Y have joint density $f_{X,Y}(x, y)$, find the joint density of X and $X + Y$.

(d) If $X_1, \ldots, X_n$ have joint density $f_{\mathbf{X}}(x_1, \ldots, x_n)$ find the joint density of $X_1, X_1 + X_2, \ldots, X_1 + \cdots + X_n$.

0.11. If X and Y are independent Poisson random variables of parameters λ and v, use the moment generating function to show $X + Y$ is Poisson of parameter $\lambda + v$. Why is $X - Y$ not a Poisson random variable?

0.12. Suppose that X is a $B_{m,p}$ random variable, Y a $B_{n,p}$ random variable, and that they are independent. Show that $X + Y$ is $B_{n+m,p}$, both directly and by using the moment generating function.

0.13. Show by using the moment generating function that if $X_1, X_2, \ldots$ are jointly normally distributed, then

(a) pairwise independence implies mutual independence.

(b) cov $(X_i, X_j) = 0$ if and only if X_i and X_j are independent.

Conclude that a family of normal random variables is independent if and only if all the covariances are zero. Also conclude that if every pair X_i, X_j is uncorrelated, then the family is mutually independent.

0.14. If X and Y are independent normal random variables, show directly, and by using the moment generating function, that $X + Y$ and $X - Y$ are

(a) normal random variables.

(b) independent.

0.15. If X and Y are independent Bernoulli random variables, that is,

$$P[X = i] = \begin{cases} \frac{1}{2}, & i = 0 \\ \frac{1}{2}, & i = 1 \\ 0, & \text{else} \end{cases}$$

show that $\frac{1}{2}(X + Y)$ and $|X - Y|$ are dependent and uncorrelated.

0.16. Suppose that X and Y are jointly normally distributed.

(a) Show that $Z = aX + bY$ is normally distributed for any values of a and b.

(b) Express the mean and variance of Z in terms of the means, variances, and covariance of the random variables X and Y.

0.17. If $X \geq 0$, show the basic inequality

$$P[X \geq a] \leq \frac{E[X]}{a}$$

and use it to demonstrate Chebyshev's inequality.

0.18. **(a)** Show that convergence in probability implies convergence in distribution.

(b) Provide a counterexample to show that the converse is not true.

0.19. **(a)** Demonstrate $a^\lambda b^{1-\lambda} \leq \lambda a + (1 - \lambda)b$ for $a, b > 0, 0 < \lambda < 1$ by showing that $(1 - \lambda) + \lambda t - t^\lambda$ achieves its maximum value of 0 at $t = 1$, and then setting $t = a/b$.

(b) Set $a = X/(E[X^p])^{1/p}$, $b = Y/(E[Y^q])^{1/q}$, $\lambda = 1/p$ in part (a) to show Holder's inequality,

$$E[|XY|] \leq (E[X^p])^{1/p}(E[Y^q])^{1/q}$$

where $p, q > 0$ and $1/p + 1/q = 1$.

0.20. Show Minkowski's inequality

$$(E[|X + Y|^p])^{1/p} \leq (E[|X|^p])^{1/p} + (E[|Y|^p])^{1/p}$$

for $p \geq 1$, by applying Holder's inequality to

$$E[|X + Y|^p] \leq E[|X + Y|^{p-1}|X|] + E[|X + Y|^{p-1}|Y|]$$

0.21. If $X_1, \ldots, X_n$ are independent identically distributed normal random variables of mean zero and variance 1, show by using the moment generating function that

$$\frac{X_1 + \cdots + X_n}{\sqrt{n}}$$

is also normal of mean zero and variance 1.

0.22. Use the moment generating function to show that

(a) $W = \dfrac{\sum\limits_{i=1}^{n} X_i - n}{\sqrt{n}}$

becomes normal of mean zero and variance 1 as $n \to \infty$, provided that the X_i are independent random variables with density $f_X(x) = e^{-x}$.

(b) $W = \dfrac{2B_n - n}{\sqrt{n}}$

also becomes normal of mean zero and variance 1, where B_n is binomial with $p = \frac{1}{2}$.

0.23. Suppose that X and Y are discrete random variables, taking only the values 0 or 1. Find necessary and sufficient conditions for X and Y to be

(a) uncorrelated.

(b) independent.

0.24. Suppose that $\{X_k\}$ are independent Bernoulli random variables with $P[X_k = 1] = p_k$. Show that

$$Y = X_1 + X_2 + \cdots + X_n$$

has mean and variance

$$E[Y] = \sum_{k=1}^{n} p_k, \quad \text{var } Y = \sum_{k=1}^{n} p_k(1 - p_k)$$

For a fixed value of $E[Y]$, show that the variance is maximum if all the p_k's are equal.

0.25. Verify the following properties of conditional expectation if X and Y are jointly continuous.

(a) $E[E[X|Y]] = E[X]$

(b) $E[YE[X|Y]] = E[E[XY|Y]] = E[XY]$

0.26. Show that convergence in mean square implies convergence in law.

0.27. If X_n converges to Y in mean square, show that there exists a subsequence converging almost surely.

CHAPTER

1

The Fourier Transform and
The Fourier Inversion Formula

One of the most important tools in the theory of probability and stochastic processes is the Fourier transform. In this chapter we give a brief review of the theory of Fourier transforms without striving for the best possible conditions. We shall also rely more upon intuition than rigor. However, here and there we interpose a rigorous derivation to give the reader a little bit of the mathematical flavor of the subject. We will develop the Fourier inversion formula using basic information from calculus.

If $f(x)$ is a function, then

$$\int_{-\infty}^{\infty} f(x)e^{i\omega x} \, dx = F(\omega) \tag{1.0.1}$$

is called the Fourier transform of $f(x)$.* If f is not absolutely integrable, the integral is interpreted as the limit, when $N \to \infty$, of the integral between $-N$ and N. We shall prove, under certain conditions to be mentioned later, that

$$\begin{aligned}
f(x) &= \frac{1}{2\pi} \int_{-\infty}^{\infty} F(\omega)e^{-i\omega x} \, d\omega \\
&= \frac{1}{2\pi} \int_{-\infty}^{\infty} \left[\int_{-\infty}^{\infty} f(y)e^{i\omega y} \, dy \right] e^{-i\omega x} \, d\omega
\end{aligned} \tag{1.0.2}$$

This is called the Fourier inversion formula. We will discuss the details of its "derivation" which show that it is true for step functions, and then indicate the

---

*Most authors define $\int_{-\infty}^{\infty} f(x)e^{-i\omega x} \, dx$ to be the Fourier transform of $f(x)$.

extension to a broader class of functions. In the course of our discussion we shall introduce the Dirac δ-function and examine some of its basic properties. The chapter concludes with an introduction to the characteristic function and some applications of the Fourier inversion formula to probability theory.

1.1   AN INTEGRAL REPRESENTATION OF THE SGN FUNCTION

In this section we verify the formula

$$\frac{1}{\pi} \int_{-\infty}^{\infty} \frac{\sin \alpha x}{x} \, dx = \text{sgn}(\alpha) = \begin{cases} 1, & \alpha > 0 \\ 0, & \alpha = 0 \\ -1, & \alpha < 0 \end{cases} \tag{1.1.1}$$

This representation plays a major role in the derivation of the Fourier inversion formula. The case $\alpha = 0$ is trivial. For $\alpha \neq 0$, if the integrals below exist, letting $\alpha x = y$ so that $\alpha \, dx = dy$, we have

$$\int_{-\infty}^{\infty} \frac{\sin \alpha x}{x} \, dx = \begin{cases} \int_{-\infty}^{\infty} \dfrac{\sin y}{y} \, dy, & \alpha > 0 \\ -\int_{-\infty}^{\infty} \dfrac{\sin y}{y} \, dy, & \alpha < 0 \end{cases}$$

$$= \text{sgn}(\alpha) \int_{-\infty}^{\infty} \frac{\sin y}{y} \, dy \tag{1.1.2}$$

and since the integrand is even

$$= 2 \, \text{sgn}(\alpha) \int_{0}^{\infty} \frac{\sin y}{y} \, dy$$

We can evaluate

$$\int_{0}^{\infty} \frac{\sin y}{y} \, dy$$

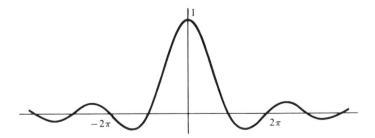

Fig. 1.1. $y = (\sin x)/x$

by **Abel's continuity theorem,** which we state without proof:

If $\int_0^\infty f(x)\,dx$ exists, then

$$\int_0^\infty f(x)\,dx = \lim_{\epsilon \to 0+} \int_0^\infty e^{-\epsilon x} f(x)\,dx \qquad (1.1.3)$$

To do this, first we must show that

$$\int_0^\infty \frac{\sin x}{x}\,dx \qquad (1.1.4)$$

exists and then we can find its value by calculating

$$\lim_{\epsilon \to 0+} \int_0^\infty e^{-\epsilon x} \frac{\sin x}{x}\,dx$$

We would have existence automatically if $(\sin x)/x$ were absolutely integrable, but unfortunately,

$$\int_0^\infty \left|\frac{\sin x}{x}\right|\,dx = \infty \qquad (1.1.5)$$

so we use another approach to show the existence of this integral. We start by writing

$$\int_0^A \frac{\sin x}{x}\,dx = \int_0^\epsilon \frac{\sin x}{x}\,dx + \int_\epsilon^A \frac{d(1 - \cos x)}{x}$$

which upon integration by parts

$$= \int_0^\epsilon \frac{\sin x}{x}\,dx + \frac{1 - \cos x}{x}\Bigg]_\epsilon^A + \int_\epsilon^A \frac{1 - \cos x}{x^2}\,dx$$

$$= \int_0^\epsilon \frac{\sin x}{x}\,dx + \frac{1 - \cos A}{A} - \frac{1 - \cos \epsilon}{\epsilon} + \int_\epsilon^A \frac{1 - \cos x}{x^2}\,dx$$

so that letting $A \to \infty$, but keeping ϵ fixed, yields

$$\int_0^\infty \frac{\sin x}{x}\,dx = \int_0^\epsilon \frac{\sin x}{x}\,dx - \frac{1 - \cos \epsilon}{\epsilon} + \int_\epsilon^\infty \frac{1 - \cos x}{x^2}\,dx$$

The first and third terms of the right-hand side are finite since

$$\left|\frac{\sin x}{x}\right| \le 1$$

and

$$0 \le \frac{1 - \cos x}{x^2} \le \frac{2}{x^2}$$

(see Fig. 1.2). We have thus proved that the integral (1.1.4) exists. Now we let

$$g(\epsilon) = \int_0^\infty e^{-\epsilon x} \frac{\sin x}{x} dx$$

and note that

$$g'(\epsilon) = -\int_0^\infty e^{-\epsilon x} \sin x \, dx = -\int_0^\infty e^{-\epsilon x} \operatorname{Im}(e^{ix}) \, dx$$

$$= -\operatorname{Im}\int_0^\infty e^{-\epsilon x} e^{ix} \, dx = -\operatorname{Im}\left(\frac{1}{\epsilon - i}\right) = -\frac{1}{\epsilon^2 + 1}$$

so that

$$\int_0^\infty e^{-\epsilon x} \frac{\sin x}{x} dx = -\tan^{-1}\epsilon + c \tag{1.1.6}$$

To evaluate c, since

$$\left|\int_0^\infty e^{-\epsilon x} \frac{\sin x}{x} dx\right| \le \int_0^\infty e^{-\epsilon x} \left|\frac{\sin x}{x}\right| dx$$

$$\le \int_0^\infty e^{-\epsilon x} dx = \frac{1}{\epsilon}$$

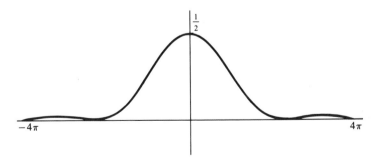

Fig. 1.2. $y = (1 - \cos x)/x^2$

let $\epsilon \to \infty$ in (1.1.6) and observe that

$$c = \frac{\pi}{2}$$

Finally, by the theorem,

$$\int_0^\infty \frac{\sin x}{x}\, dx = \lim_{\epsilon \to 0+} \int_0^\infty e^{-\epsilon x} \frac{\sin x}{x}\, dx$$

$$= \lim_{\epsilon \to 0+} \tan^{-1}\epsilon + \frac{\pi}{2}$$

$$= \frac{\pi}{2}$$

which by (1.1.2) completes the proof of (1.1.1),

$$\frac{1}{\pi} \int_{-\infty}^\infty \frac{\sin \alpha x}{x}\, dx = \text{sgn}\,(\alpha) \qquad\qquad (1.1.7)$$

1.2   FOURIER INVERSION OF UNIT RECTANGLE FUNCTIONS

We now use the results of the preceding section to verify the Fourier inversion formula for unit rectangle functions. We call

$$f_\alpha(x) = \tfrac{1}{2}[\text{sgn}\,(\alpha + x) + \text{sgn}\,(\alpha - x)] \qquad\qquad (1.2.1)$$

a unit rectangle function centered at 0 of width 2α (see Fig. 1.3). By (1.1.7)

$$f_\alpha(x) = \frac{1}{2}\left[\frac{1}{\pi} \int_{-\infty}^\infty \frac{\sin (\alpha + x)\omega}{\omega}\, d\omega + \frac{1}{\pi} \int_{-\infty}^\infty \frac{\sin (\alpha - x)\omega}{\omega}\, d\omega \right]$$

$$= \frac{1}{2\pi} \int_{-\infty}^\infty \frac{\sin (\alpha + x)\omega + \sin (\alpha - x)\omega}{\omega}\, d\omega$$

$$= \frac{1}{\pi} \int_{-\infty}^\infty \frac{\sin \alpha\omega}{\omega} \cos \omega x\, d\omega$$

Noting that

$$\frac{1}{\pi} \int_{-\infty}^\infty \frac{\sin \alpha\omega}{\omega} \sin \omega x\, d\omega = 0$$

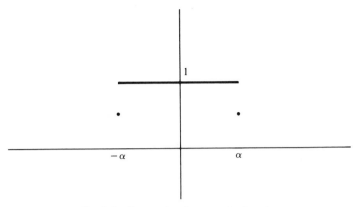

Fig. 1.3. Centered unit rectangle function

since its integrand is odd, we conclude that

$$f_\alpha(x) = \frac{1}{\pi} \int_{-\infty}^{\infty} \frac{\sin \alpha\omega}{\omega} e^{-i\omega x} \, d\omega \tag{1.2.2}$$

Clearly,

$$\frac{\sin \alpha\omega}{\omega} = \frac{1}{2} \int_{-\alpha}^{\alpha} e^{i\omega y} \, dy = \frac{1}{2} \int_{-\infty}^{\infty} f_\alpha(y) e^{i\omega y} \, dy$$

and substituting this into (1.2.2) gives

$$f_\alpha(x) = \frac{1}{2\pi} \int_{-\infty}^{\infty} \left[\int_{-\infty}^{\infty} f_\alpha(y) e^{i\omega y} \, dy \right] e^{-i\omega x} \, d\omega \tag{1.2.3}$$

This is the Fourier inversion formula for unit rectangle functions centered at 0. More generally, replacing y by $y - \beta$ and x by $x - \beta$ shows that the Fourier inversion formula holds for unit rectangle functions $f_{\alpha\beta}(y) = f_\alpha(y - \beta)$ of width 2α centered at β (see Fig. 1.4).

$$f_{\alpha\beta}(x) = \frac{1}{2\pi} \int_{-\infty}^{\infty} \left[\int_{-\infty}^{\infty} f_{\alpha\beta}(y) e^{i\omega y} \, dy \right] e^{-i\omega x} \, d\omega \tag{1.2.4}$$

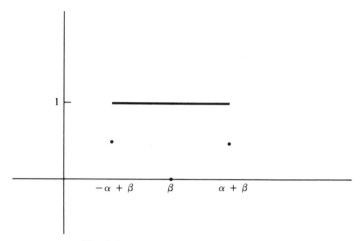

Fig. 1.4. Arbitrary unit rectangle function

1.3   STEP FUNCTIONS AND BEYOND, APPLICATION TO DENSITY FUNCTIONS

We call $f(x)$ a step function if it is a finite linear combination

$$f(x) = \sum_{i=1}^{n} c_i f_{\alpha_i \beta_i} \tag{1.3.1}$$

of unit rectangle functions. Since (1.2.4) is linear in $f_{\alpha\beta}$, it holds for any step function. It seems reasonable that if an arbitrary function g is sufficiently strongly approximable by step functions, the Fourier inversion formula should still hold. This is beyond the scope of this book; however, it is shown at the end of Chapter 2 that if f is both absolutely and square integrable, then

$$h(x) = \frac{1}{2\pi} \int_{-\infty}^{\infty} \left[\int_{-\infty}^{\infty} f(y) e^{i\omega y} \, dy \right] e^{-i\omega x} \, d\omega$$

satisfies

$$\int_{-\infty}^{\infty} [f(x) - h(x)]^2 \, dx = 0$$

This implies that

$$\int_{-\infty}^{\infty} |f(x) - h(x)| = 0$$

so if f is a density function, h and f are density functions from the same distribution, that is,

$$\int_{-\infty}^{x} h(t)\, dt = \int_{-\infty}^{x} f(t)\, dt$$

Since all probability densities are absolutely integrable, and nearly all those of interest are square integrable, the Fourier inversion formula holds in this sense in a wide variety of situations.

1.4   A DIGRESSION ABOUT THE DIRAC δ-FUNCTION

We now discuss the Dirac δ-function and give some of its important properties, based on the Fourier inversion formula. Looking at this formula and changing the order of integration we find, formally,

$$
\begin{aligned}
f(x) &= \frac{1}{2\pi} \int_{-\infty}^{\infty} \left[\int_{-\infty}^{\infty} f(y) e^{i\omega y}\, dy \right] e^{-i\omega x}\, d\omega \\
&= \int_{-\infty}^{\infty} f(y) \left[\frac{1}{2\pi} \int_{-\infty}^{\infty} e^{i\omega y} e^{-i\omega x}\, d\omega \right] dy \qquad (1.4.1) \\
&= \int_{-\infty}^{\infty} f(y) \left[\frac{1}{2\pi} \int_{-\infty}^{\infty} e^{i\omega(y-x)}\, d\omega \right] dy
\end{aligned}
$$

If we ignore the fact that the inside integral does not exist, and define the Dirac δ-function by

$$\delta(y - x) = \frac{1}{2\pi} \int_{-\infty}^{\infty} e^{i\omega(y-x)}\, d\omega \qquad (1.4.2)$$

then (1.4.1) becomes

$$f(x) = \int_{-\infty}^{\infty} f(y)\delta(y - x)\, dy \qquad (1.4.3)$$

Since the change of variables $\omega' = -\omega$ yields

$$\delta(y - x) = \delta(x - y)$$

we can also write (1.4.1) in the form

$$f(x) = \int_{-\infty}^{\infty} f(y)\delta(x - y)\, dy \qquad (1.4.4)$$

This formula has a simple interpretation. For a function $k(x, y)$, an integral of the form

$$\int_{-\infty}^{\infty} k(x, y)f(y)\, dy$$

is a linear operator acting on f, and can be regarded as an infinite-dimensional analogue of the finite-dimensional action of matrix $\mathbf{K} = (k_{ij})$ on a vector $\mathbf{f} = (f_i)$:

$$\mathbf{Kf} = \sum_{j=1}^{n} k_{ij} f_j$$

For matrices it is easy to write down the unit matrix $\mathbf{I}$, which for all $\mathbf{f}$ satisfies

$$\mathbf{f} = \mathbf{If}$$

As an operator, $\mathbf{I}$ leaves the vector unchanged. One can consider δ as the infinite dimensional analogue of this unit matrix, since for any function f to which the Fourier inversion formula applies,

$$f(x) = \int_{-\infty}^{\infty} f(y)\delta(x - y)\, dy$$

As an operator, integrating against δ leaves the function unchanged, which indicates that it is the "unit operator." This analogy is even more appealing if we use the standard notation for the convolution of two functions f and g,

$$(f * g)(x) = \int_{-\infty}^{\infty} f(y)g(x - y)\, dy$$

and write (1.4.4) as

$$f(x) = (f * \delta)(x)$$

or briefly

$$f = f * \delta \qquad\qquad (1.4.5)$$

Using (1.4.3) one can verify formally several formulas involving δ. For example, setting

$$f(x) = e^{i\omega x}$$

results in

$$e^{i\omega x} = \int_{-\infty}^{\infty} e^{i\omega y}\, \delta(y - x)\, dy$$

which when $x = 0$ says that

$$1 = \int_{-\infty}^{\infty} e^{i\omega y}\, \delta(y)\, dy \tag{1.4.6}$$

is the Fourier transform of δ. Also, applying (1.4.3) to the unit rectangle function f_ϵ gives

$$f_\epsilon(x) = \int_{-\infty}^{\infty} f_\epsilon(y)\delta(y - x)\, dy$$

which, when $x = 0$, shows that δ is a probability density concentrated on $(-\epsilon, \epsilon)$:

$$1 = \int_{-\epsilon}^{\epsilon} \delta(y)\, dy \tag{1.4.7}$$

Finally, if we let $f(x)$ be the function $h_t(x)$ given by

$$h_t(x) = \begin{cases} 1, & x \le t \\ 0, & x > t \end{cases} \tag{1.4.8}$$

then

$$h_t(x) = \int_{-\infty}^{\infty} h_t(y)\, \delta(y - x)\, dy = \int_{-\infty}^{t} \delta(y - x)\, dy$$

which if $x = 0$ is

$$h_t(0) = \int_{-\infty}^{t} \delta(y)\, dy$$

Since as a function of t, $h_t(0)$ is just the Heaviside step function $H(t)$,

$$H(t) = \begin{cases} 0, & t < 0 \\ 1, & t \ge 0 \end{cases} \tag{1.4.9}$$

from this we obtain, formally,

$$\delta(t) = \frac{dH(t)}{dt} \tag{1.4.10}$$

implying that

$$\delta(t) = 0, \qquad t \ne 0$$

This gives us a working definition of $\delta(t)$ as a probability density "function" that is zero, except at $t = 0$, and satisfies (1.4.10). It is possible to make these statements rigorous by introducing the concept of generalized functions, but as this belongs more to the field of functional analysis than to probability theory, we shall not go into that.

1.5   AN EXAMPLE OF A FOURIER INVERSION

Consider the function

$$f(x) = \tfrac{1}{2}e^{-|x|}$$

Its Fourier transform is given by

$$F(\omega) = \int_{-\infty}^{\infty} \tfrac{1}{2}e^{-|x|}e^{i\omega x}\, dx \tag{1.5.1}$$

which since the complex part of the integrand is odd, and the real part is even,

$$= \operatorname{Re} \int_0^\infty e^{-x}e^{i\omega x}\, dx$$

$$= \operatorname{Re} \left(\frac{1}{1 - i\omega} \right)$$

$$= \frac{1}{1 + \omega^2}$$

We now verify the Fourier inversion formula for this special case. In other words, we show that

$$\frac{1}{2}\, e^{-|x|} = \frac{1}{2\pi} \int_{-\infty}^{\infty} \frac{1}{1 + \omega^2}\, e^{-i\omega x}\, d\omega \tag{1.5.2}$$

To do this we use a formal method which demonstrates the use of the δ-function. Let

$$G(x) = \frac{1}{2\pi} \int_{-\infty}^{\infty} \frac{1}{1 + \omega^2}\, e^{-i\omega x}\, d\omega$$

which is a bounded continuous real valued function for all x. Formally differentiating G two times with respect to x yields

$$G''(x) = \frac{1}{2\pi} \int_{-\infty}^{\infty} \frac{-\omega^2}{1 + \omega^2}\, e^{-i\omega x}\, d\omega$$

so we have

$$G(x) - G''(x) = \frac{1}{2\pi} \int_{-\infty}^{\infty} e^{-i\omega x} \, d\omega = \delta(x)$$

Applying the properties of δ to this results in the following three statements:

$$G(x) - G''(x) = 0, \qquad x < 0 \tag{1.5.3}$$

$$\int_{-\epsilon}^{\epsilon} (G(x) - G''(x)) \, dx = 1, \qquad \epsilon > 0 \tag{1.5.4}$$

$$G(x) - G''(x) = 0, \qquad x > 0 \tag{1.5.5}$$

Solving the differential equations (1.5.3) and (1.5.5), since $G(x)$ is bounded as $x \to \pm\infty$, yields

$$G(x) = \begin{cases} ae^x, & x < 0 \\ be^{-x}, & x > 0 \end{cases} \tag{1.5.6}$$

Since G is continuous at $x = 0$,

$$a = b \tag{1.5.7}$$

We can apply the fundamental theorem of the calculus to (1.5.4) and get

$$\int_{-\epsilon}^{\epsilon} G(x) \, dx - G'(\epsilon) + G'(-\epsilon) = 1$$

Letting $\epsilon \to 0$, since G is bounded, we conclude that

$$G'(0-) - G'(0+) = 1$$

or

$$a + b = 1$$

This and (1.5.7) gives $a = b = \frac{1}{2}$, so by (1.5.6)

$$G(x) = \tfrac{1}{2} e^{-|x|} \tag{1.5.8}$$

as required.

1.6   THE MULTIDIMENSIONAL FOURIER TRANSFORM AND INVERSION FORMULA

We now state the higher-dimensional analogue of the Fourier transform and inversion formula, for future reference. If $f(x_1, \ldots, x_n)$ is a real-valued function, we define its Fourier transform $F(\xi_1, \ldots, \xi_n)$ by

$$F(\xi_1, \ldots, \xi_n) = \int_{-\infty}^{\infty} \cdots \int_{-\infty}^{\infty} f(x_1, \ldots, x_n) e^{i(\xi_1 x_1 + \cdots + \xi_n x_n)} \, dx_1 \cdots dx_n$$

(1.6.1)

The corresponding inversion formula is

$$f(x_1, \ldots, x_n) = \left(\frac{1}{2\pi}\right)^n \int_{-\infty}^{\infty} \cdots \int_{-\infty}^{\infty} F(\xi_1, \ldots, \xi_n) e^{-i(\xi_1 x_1 + \cdots + \xi_n x_n)} \, d\xi_1 \cdots d\xi_n$$

(1.6.2)

In terms of the vectors $\mathbf{x} = (x_1, \ldots, x_n)^{\mathrm{T}}$ and $\boldsymbol{\xi} = (\xi_1, \ldots, \xi_n)^{\mathrm{T}}$, these can be written in the condensed forms

$$F(\boldsymbol{\xi}) = \int_{-\infty}^{\infty} \cdots \int_{-\infty}^{\infty} f(\mathbf{x}) e^{i(\boldsymbol{\xi}^{\mathrm{T}}\mathbf{x})} \, dx_1 \cdots dx_n$$

(1.6.3)

and

$$f(\mathbf{x}) = \left(\frac{1}{2\pi}\right)^n \int_{-\infty}^{\infty} \cdots \int_{-\infty}^{\infty} F(\boldsymbol{\xi}) e^{-i(\boldsymbol{\xi}^{\mathrm{T}}\mathbf{x})} \, d\xi_1 \cdots d\xi_n$$

(1.6.4)

1.7   CHARACTERISTIC FUNCTIONS: APPLICATIONS OF THE FOURIER INVERSION FORMULA TO PROBABILITY THEORY

Let X be a random variable. The characteristic function C_X of X is defined by

$$C_X(\xi) = E[e^{i\xi X}]$$

(1.7.1)

If X has a density function $f_X(x)$, the characteristic function of X has the form

$$C_X(\xi) = \int_{-\infty}^{\infty} f_X(x) e^{i\xi x} \, dx$$

(1.7.2)

which is the Fourier transform of the density function. If X is a discrete random variable with mass function p_X, its characteristic function is

$$C_X(\xi) = \sum_{x_j} p_X(x_j)e^{i\xi x_j} \qquad (1.7.3)$$

The Fourier inversion formula applied to characteristic functions is

$$f_X(x) = \frac{1}{2\pi} \int_{-\infty}^{\infty} C_X(\xi)e^{-i\xi x} \, d\xi \qquad (1.7.4)$$

Finding the characteristic function C_X and then using the Fourier inversion formula to recover the density function f_X is one of the standard methods in probability theory for showing independence or for calculating the density of sums of random variables. Also, we shall see that the moments of a random variable can be found from the derivatives of its characteristic function. Recall that in Chapter 0 moment generating functions were used for the same purposes. However, the Fourier inversion formula makes characteristic functions a better tool than moment generating functions, and they have the further advantage of being defined for any density function and all values of ξ. Thus they may be applied in situations where moment generating functions cannot. We now present some examples.

Discrete random variables

If X is a discrete random variable, its distribution function will have an upward step at each possible value of X, and we shall see that the related formal probability density function is a sum of Dirac δ-functions. This conclusion is suggested by the fact that the distribution function of X can be considered as a linear combination

$$\sum_{i=1}^{\infty} p_i H(x - x_i)$$

of Heaviside step functions. Since the density function is the derivative of the distribution function, the relation (1.4.10)

$$\delta(x) = \frac{dH(x)}{dx}$$

implies formally that the density function is

$$\frac{d}{dx} \sum_{i=1}^{\infty} p_i H(x - x_i) = \sum_{i=1}^{\infty} p_i \, \delta(x - x_i)$$

We may also demonstrate this fact by using characteristic functions. Given a discrete random variable X satisfying

$$P[X = x_j] = p_j, \quad j = 1, 2, \ldots, \quad \sum_{j=1}^{\infty} p_j = 1$$

its characteristic function is

$$C_X(\xi) = E[e^{i\xi X}] = p_1 e^{i\xi x_1} + p_2 e^{i\xi x_2} + \cdots$$

Applying the inversion (1.7.4) yields

$$f_X(x) = \frac{1}{2\pi} \int_{-\infty}^{\infty} (p_1 e^{i\xi x_1} + p_2 e^{i\xi x_2} + \cdots) e^{-i\xi x} \, d\xi$$

$$= \frac{1}{2\pi} \int_{-\infty}^{\infty} p_1 e^{i\xi(x_1 - x)} \, d\xi + \frac{1}{2\pi} \int_{-\infty}^{\infty} p_2 e^{i\xi(x_2 - x)} \, d\xi + \cdots$$

which by the definition of the δ function,

$$= p_1 \delta(x - x_1) + p_2 \delta(x - x_2) + \cdots$$

as expected. We also note that this gives

$$P[X \le x] = \int_{-\infty}^{x} \sum_{k=1}^{\infty} p_k \delta(t - x_k) \, dt$$

which by (1.4.10)

$$= \sum_{k=1}^{\infty} p_k H(x - x_k)$$

$$= \sum_{k=1}^{\infty} P[X = x_k] H(x - x_k)$$

This, by the definition of the Heaviside function, is just the sum over all k's such that $x_k \le x$.

Sums of independent Cauchy random variables

Let $Y_1, Y_2, \ldots, Y_n$ be independent identically distributed random variables with the Cauchy density

$$f_Y(y) = \frac{1}{\pi} \frac{1}{1 + y^2}$$

and consider the average

$$Z = \frac{Y_1 + Y_2 + \cdots + Y_n}{n}$$

An elementary way to try to find the distribution of Z is to calculate its moment generating function. However, the moment generating function does not exist in this case, since all the even moments are infinite. Therefore, we try the characteristic function:

$$C_Z(\xi) = E[e^{i\xi Z}] = E\left[\exp\left(i\xi \frac{Y_1 + Y_2 + \cdots + Y_n}{n} \right) \right]$$

$$= E\left[\exp\left(i\xi \frac{Y_1}{n} \right) \cdots \exp\left(i\xi \frac{Y_n}{n} \right) \right]$$

by independence,

$$= E\left[\exp\left(i\xi \frac{Y_1}{n} \right) \right] \cdots E\left[\exp\left(i\xi \frac{Y_n}{n} \right) \right]$$

and since the X_i's are identically distributed,

$$= \left(E\left[\exp\left(i\xi \frac{Y_1}{n} \right) \right] \right)^n$$

$$= \left[\frac{1}{\pi} \int_{-\infty}^{\infty} \exp\left(\frac{i\xi y}{n} \right) \frac{1}{1 + y^2} \, dy \right]^n$$

or, setting $x = -\xi/n$, $\omega = y$ in (1.5.2)

$$= \left[\exp\left(-\frac{|\xi|}{n} \right) \right]^n = e^{-|\xi|}$$

Applying the inversion (1.7.4)

$$f_Z(z) = \frac{1}{2\pi} \int_{-\infty}^{\infty} e^{-|\xi|} e^{-i\xi z} \, d\xi$$

which by a change of variables in (1.5.1)

$$= \frac{1}{\pi} \frac{1}{1 + z^2}$$

indicating that the average of n independent identical Cauchy random variables has the same Cauchy distribution. It is instructive to note that this result holds true as $n \to \infty$, and to compare this with the law of large numbers.

The density for the sum of two independent random variables

Let X and Y be independent random variables having f_X and f_Y as their respective densities. We can find the density function for the random variable

$$Z = X + Y$$

by using its characteristic function:

$$C_Z(\xi) = E[e^{i\xi Z}]$$
$$= E[e^{i\xi(X+Y)}]$$

which by independence

$$= E[e^{i\xi X}]E[e^{i\xi Y}]$$
$$= \int_{-\infty}^{\infty} e^{i\xi x} f_X(x)\, dx \int_{-\infty}^{\infty} e^{i\xi y} f_Y(y)\, dy$$

or by the change of variables $w = x + y$

$$= \int_{-\infty}^{\infty} e^{i\xi w} \int_{-\infty}^{\infty} f_X(x) f_Y(w - x)\, dx\, dw$$

The inversion formula (1.7.4) in this case gives

$$f_Z(z) = \frac{1}{2\pi} \int_{-\infty}^{\infty} C_Z(\xi) e^{-i\xi z}\, d\xi$$
$$= \frac{1}{2\pi} \int_{-\infty}^{\infty} \left[\int_{-\infty}^{\infty} e^{i\xi w} \int_{-\infty}^{\infty} f_X(x) f_Y(w - x)\, dx\, dw \right] e^{-i\xi z}\, d\xi$$
$$= \frac{1}{2\pi} \int_{-\infty}^{\infty} \left[\int_{-\infty}^{\infty} \left[\int_{-\infty}^{\infty} f_X(x) f_Y(w - x)\, dx \right] e^{i\xi w}\, dw \right] e^{-i\xi z}\, d\xi$$

which is the Fourier inversion formula applied to the innermost integral, so by Section 1.3,

$$f_Z(z) = \int_{-\infty}^{\infty} f_X(x) f_Y(z - x)\, dx$$

or in short,

$$f_Z = f_X * f_Y \qquad (1.7.5)$$

Characteristic function of a normal random variable

If X is a normal random variable with density

$$f_X(x) = \frac{1}{\sqrt{2\pi}} e^{-(1/2)x^2}$$

then

$$C_X(\xi) = \int_{-\infty}^{\infty} e^{i\xi x} \frac{1}{\sqrt{2\pi}} e^{-(1/2)x^2} \, dx$$

To evaluate this integral, we write

$$\frac{d}{d\xi} C_X(\xi) = \int_{-\infty}^{\infty} i x e^{i\xi x} \frac{1}{\sqrt{2\pi}} e^{-(1/2)x^2} \, dx$$

and integrate by parts

$$= -\xi \int_{-\infty}^{\infty} e^{i\xi x} \frac{1}{\sqrt{2\pi}} e^{-(1/2)x^2} \, dx$$

$$= -\xi C_X(\xi)$$

The solution to this differential equation with the obvious initial condition

$$C_X(0) = 1$$

is

$$C_X(\xi) = e^{-(1/2)\xi^2} \qquad (1.7.6)$$

By a change of variables, this can be used to show that a normal random variable with mean μ and variance σ^2 has characteristic function

$$C_X(\xi) = \exp\left(i\mu\xi - \tfrac{1}{2}\sigma^2\xi^2\right) \qquad (1.7.7)$$

Moments of random variables

The moments of a random variable X can be obtained by differentiating its characteristic function $C_X(\xi)$ with respect to ξ and then setting $\xi = 0$,

$$\frac{d^k}{d\xi^k} C_X(\xi)\bigg|_{\xi=0} = \frac{d^k}{d\xi^k} E[e^{i\xi X}]\bigg|_{\xi=0}$$

$$= E[(iX)^k e^{i\xi X}]\bigg|_{\xi=0}$$

$$= i^k E[X^k]$$

or

$$E[X^k] = \frac{1}{i^k} \frac{d^k}{d\xi^k} C_X(\xi)\bigg|_{\xi=0} \tag{1.7.8}$$

As an application of this, we find the moments of a normal random variable X of mean zero and variance 1:

$$E[X] = \frac{1}{i} \frac{d}{d\xi} \exp\left(-\frac{1}{2}\xi^2\right)\bigg|_{\xi=0} = 0$$

$$E[X^2] = \frac{1}{i^2} \frac{d^2}{d\xi^2} \exp\left(-\frac{1}{2}\xi^2\right)\bigg|_{\xi=0} = 1$$

$$E[X^3] = \frac{1}{i^3} \frac{d^3}{d\xi^3} \exp\left(-\frac{1}{2}\xi^2\right)\bigg|_{\xi=0} = 0$$

$$E[X^4] = \frac{1}{i^4} \frac{d^4}{d\xi^4} \exp\left(-\frac{1}{2}\xi^2\right)\bigg|_{\xi=0} = 3$$

Properties of characteristic functions, Bochner's theorem

From the definition of the characteristic function of a random variable X, we have the following:

(i) $C_X(\xi)$ has the symmetry property $\overline{C_X(\xi)} = C_X(-\xi)$.
(ii) $C_X(\xi)$ is continuous, in particular it is continuous at 0. This means that $C_X(\xi) \to 1$ as $\xi \to 0$. This is because

$$C_X(\xi + \Delta\xi) - C_X(\xi) = E[e^{i(\xi + \Delta\xi)X} - e^{i\xi X}]$$

so that

$$|C_X(\xi + \Delta\xi) - C_X(\xi)| \le E[|e^{i(\xi + \Delta\xi)X} - e^{i\xi X}|] \le E[|e^{i\Delta\xi X} - 1|]$$

which converges to zero by the bounded convergence theorem. In fact, this shows that $C_X(\xi) \to 1$ as $\xi \to 0$ and that C_X is uniformly continuous as well.

(iii) $C_X(\xi)$ is positive definite, meaning that for any n, complex numbers z_i, and real ξ_i, $1 \le i \le n$,

$$\sum_{i=1}^{n} \sum_{j=1}^{n} C_X(\xi_i - \xi_j) z_i \bar{z}_j \ge 0$$

This is because

$$0 \le E\left[\left|\sum_{j=1}^{n} e^{i\xi_j X} z_j\right|^2\right] = \sum_{i=1}^{n} \sum_{j=1}^{n} C_X(\xi_i - \xi_j) z_i \bar{z}_j$$

We note that positive definiteness implies (i) above.

Bochner's theorem says that the positive definite functions $C(\xi)$ which are continuous at 0 and satisfy $C(0) = 1$, are exactly the characteristic functions. We shall not prove this here. Thus the characteristic functions can be completely characterized by properties that do not depend on probability theory. If the corresponding random variable has the density function f, then

$$C(\xi) = \int_{-\infty}^{\infty} e^{i\xi t} f(t) \, dt$$

More generally, the continuous positive definite functions $C(\xi)$ are those which are Stieltjes transforms of bounded nondecreasing functions F:

$$C(\xi) = \int_{-\infty}^{\infty} e^{i\xi t} \, dF(t)$$

$C(0) = 1$ is obtained when F is a probability distribution function.

The continuity theorem, central limit theorem

One of the most important properties of the characteristic function is given by the continuity theorem, which we state without proof. Let $X_1, X_2, \ldots$ be a sequence of random variables, and $C_{X_1}, C_{X_2}, \ldots$ the corresponding characteristic functions. We have the following:

(i)   If $X_n \to X$ in law, then $\lim_{n \to \infty} C_{X_n}(\xi) = C_X(\xi)$ for each ξ.

(ii)   If $C(\xi) = \lim\limits_{n \to \infty} C_{X_n}(\xi)$ exists and is continuous at $\xi = 0$, then there

exists a random variable X so that $C(\xi) = C_X(\xi)$ and $X_n \to X$ in law.

As an example of the use of this property, we can give a proof of the **central limit theorem.** Suppose that $X_1, X_2, \ldots$ are independent identically distributed random variables of mean μ and variance $\sigma^2 > 0$. The central limit theorem states that

$$W_n = \frac{\sum\limits_{i=1}^{n} X_i - n\mu}{\sigma\sqrt{n}}$$

converges in law to a normal distribution of mean zero and variance 1. Noting that we can assume the X_i's have mean zero, we calculate the characteristic function of W_n

$$C_{W_n}(\xi) = E[e^{i\xi W_n}]$$

$$= E\left[\exp\left(i\xi \frac{\sum\limits_{i=1}^{n} X_i}{\sigma\sqrt{n}}\right)\right]$$

by independence

$$= \left[C_X\left(\frac{\xi}{\sigma\sqrt{n}}\right)\right]^n$$

and by the Taylor series expansion of $C_X(\xi)$

$$= \left[1 + \frac{i^2\sigma^2}{2}\left(\frac{\xi}{\sigma\sqrt{n}}\right)^2 + o\left(\frac{\xi}{\sigma\sqrt{n}}\right)^2\right]^n$$

$$= \left[1 - \frac{\xi^2}{2n} + o\left(\frac{\xi}{\sigma\sqrt{n}}\right)^2\right]^n$$

so

$$\lim\limits_{n \to \infty} C_{W_n}(\xi) = e^{-(1/2)\xi^2}$$

Applying the continuity theorem now completes the proof.

1.8 MULTIDIMENSIONAL CHARACTERISTIC FUNCTION AND ITS INVERSION FORMULA

If $\mathbf{X} = (X_1, \ldots, X_n)^T$ is a vector-valued random variable with the density $f_{\mathbf{X}}$, its characteristic function, for $\boldsymbol{\xi} = (\xi_1, \ldots, \xi_n)^T$, is defined by

$$
\begin{aligned}
C_{\mathbf{X}}(\boldsymbol{\xi}) &= E[e^{i\boldsymbol{\xi}^T\mathbf{X}}] \\
&= E\left[\exp\left(i\sum_{j=1}^{n}\xi_j X_j\right)\right] \\
&= \int_{-\infty}^{\infty}\cdots\int_{-\infty}^{\infty}\exp\left(i\sum_{j=1}^{n}\xi_j x_j\right)f_{\mathbf{X}}(x_1, \ldots, x_n)\,dx_1\cdots dx_n
\end{aligned}
\tag{1.8.1}
$$

which is the Fourier transform of the density. The corresponding inversion formula is

$$
f_{\mathbf{X}}(\mathbf{x}) = \left(\frac{1}{2\pi}\right)^n\int_{-\infty}^{\infty}\cdots\int_{-\infty}^{\infty}C_{\mathbf{X}}(\boldsymbol{\xi})e^{-i\boldsymbol{\xi}^T\mathbf{x}}\,d\xi_1\cdots d\xi_n
\tag{1.8.2}
$$

We use this in the following important example.

Independence of random variables

If X and Y are independent random variables, with densities f_X and f_Y, respectively, then

$$
C_{X,Y}(\xi, \eta) = C_X(\xi)C_Y(\eta)
\tag{1.8.3}
$$

The converse is also true, that is, if (1.8.3) holds for all real ξ and η, then X and Y are independent. The proof is a straightforward application of the multidimensional Fourier inversion formula.

Let $f_{X,Y}$ be the joint density of X and Y, so by (1.8.2) we can write

$$
f_{X,Y}(x, y) = \left(\frac{1}{2\pi}\right)^2\int_{-\infty}^{\infty}\int_{-\infty}^{\infty}C_{X,Y}(\xi, \eta)e^{-i(\xi x+\eta y)}\,d\xi\,d\eta
$$

and by (1.8.3)

$$
= \left(\frac{1}{2\pi}\right)^2\int_{-\infty}^{\infty}\int_{-\infty}^{\infty}C_X(\xi)C_Y(\eta)e^{-i(\xi x+\eta y)}\,d\xi\,d\eta
$$

or

$$= \left[\frac{1}{2\pi}\int_{-\infty}^{\infty} C_X(\xi)e^{-i\xi x}\, d\xi\right]\left[\frac{1}{2\pi}\int_{-\infty}^{\infty} C_Y(\eta)e^{-i\eta y}\, d\eta\right]$$

which by the inversion formula (1.7.4) gives

$$f_{X,Y}(x,\, y) = f_X(x)f_Y(y)$$

as required.

More generally, $X_1, \ldots, X_n$ are independent random variables if and only if

$$E\left[\exp\left(i\sum_{k=1}^{n} \xi_k X_k\right)\right] = \prod_{k=1}^{n} E[\exp(i\xi_k X_k)] \qquad (1.8.4)$$

or

$$C_{\mathbf{X}}(\boldsymbol{\xi}) = \prod_{k=1}^{n} C_{X_k}(\xi_k) \qquad (1.8.5)$$

Application to normal random variables

Suppose that X and Y are independent normal random variables of mean zero and variance 1. We can show that for every θ, $X\cos\theta + Y\sin\theta$ and $-X\sin\theta + Y\cos\theta$ are also independent normal random variables of mean zero and variance 1. By direct calculation, the joint characteristic function of these random variables is

$$E[e^{i(\xi(X\cos\theta + Y\sin\theta) + \eta(-X\sin\theta + Y\cos\theta))}]$$

$$= E[e^{i(X(\xi\cos\theta - \eta\sin\theta) + Y(\xi\sin\theta + \eta\cos\theta))}]$$

or by independence

$$= E[e^{iX(\xi\cos\theta - \eta\sin\theta)}]E[e^{iY(\xi\sin\theta + \eta\cos\theta)}]$$

and by normality

$$= \exp[-\tfrac{1}{2}(\xi\cos\theta - \eta\sin\theta)^2]\exp[-\tfrac{1}{2}(\xi\sin\theta + \eta\cos\theta)^2]$$

$$= \exp[-\tfrac{1}{2}\xi^2]\exp[-\tfrac{1}{2}\eta^2]$$

which completes the result.

Setting $\theta = 0$ gives the obvious result that X and Y are independent. Setting $\theta = -\pi/4$ gives the result that

$$\frac{X}{\sqrt{2}} + \frac{Y}{\sqrt{2}} \quad \text{and} \quad \frac{X}{\sqrt{2}} - \frac{Y}{\sqrt{2}}$$

are independent normal random variables, and this means that $X + Y$ and $X - Y$ are independent normal random variables. A partial converse to this is true, which you can find in Problem 1.10.

Functions of independent random variables

Suppose that X and Y are independent random variables, with densities f_X and f_Y. Let $A(x)$ and $B(x)$ be real-valued functions. We show that the random variables $A(X)$ and $B(Y)$ are independent. To do this we use (1.8.3), and verify that the joint characteristic function satisfies

$$C_{A(X),B(Y)}(\xi, \eta) = C_{A(X)}(\xi)C_{B(Y)}(\eta)$$

By definition,

$$C_{A(X),B(Y)}(\xi, \eta) = E[e^{i(\xi A(X) + \eta B(Y))}]$$

which by independence

$$= \int_{-\infty}^{\infty} \int_{-\infty}^{\infty} e^{i(\xi A(x) + \eta B(y))} f_X(x) f_Y(y) \, dx \, dy$$

$$= \int_{-\infty}^{\infty} e^{i\xi A(x)} f_X(x) \, dx \int_{-\infty}^{\infty} e^{i\eta B(y)} f_Y(y) \, dy$$

and by (1.7.2)

$$= C_{A(X)}(\xi)C_{B(Y)}(\eta)$$

Independence of families of random variables

Equation (1.8.3) can be extended to families of random variables. The family $(X_1, \ldots, X_k)$ will be independent of the family $(Y_1, \ldots, Y_n)$ if and only if

$$E\left[\exp\left(i\left(\sum_{j=1}^{k} \xi_j X_j + \sum_{j=1}^{n} \eta_j Y_j\right)\right)\right] = E\left[\exp\left(i\sum_{j=1}^{k} \xi_j X_j\right)\right] E\left[\exp\left(i\sum_{j=1}^{n} \eta_j Y_j\right)\right]$$

$$(1.8.6)$$

As an example, we shall show that if $X_1, \ldots, X_n$ are independent identically distributed normal random variables, mean zero and variance 1, then

$$\overline{X} = \frac{X_1 + \cdots + X_n}{n}$$

is independent of the group $X_1 - \bar{X}, \ldots, X_n - \bar{X}$. We have only to calculate

$$E\left[\exp\left(i\left(\eta\bar{X} + \sum_{k=1}^{n} \xi_k(X_k - \bar{X})\right)\right)\right]$$

$$= E\left[\exp\left(i \sum_{k=1}^{n} \left(X_k\left(\xi_k - \frac{1}{n}\sum_{j=1}^{n} \xi_j + \frac{\eta}{n}\right)\right)\right)\right]$$

using the independence of the X_k's

$$= \prod_{k=1}^{n} E\left[\exp\left(iX_k\left(\xi_k - \frac{1}{n}\sum_{j=1}^{n} \xi_j + \frac{\eta}{n}\right)\right)\right]$$

using the characteristic function of a normal random variable

$$= \prod_{k=1}^{n} \exp\left[-\frac{1}{2}\left(\xi_k - \frac{1}{n}\sum_{j=1}^{n} \xi_j + \frac{\eta}{n}\right)^2\right]$$

$$= \exp\left[-\frac{1}{2}\sum_{k=1}^{n}\left(\xi_k - \frac{1}{n}\sum_{j=1}^{n} \xi_j + \frac{\eta}{n}\right)^2\right]$$

and noting that the sum of the "cross terms" is zero

$$= \exp\left[-\frac{1}{2}\sum_{k=1}^{n}\left(\left(\xi_k - \frac{1}{n}\sum_{j=1}^{n} \xi_j\right)^2 + \left(\frac{\eta}{n}\right)^2\right)\right]$$

$$= \exp\left[-\frac{1}{2}\sum_{k=1}^{n}\left(\xi_k - \frac{1}{n}\sum_{j=1}^{n} \xi_j\right)^2\right]\exp\left(-\frac{1}{2}\frac{\eta^2}{n}\right)$$

$$= E\left[\exp\left(i \sum_{k=1}^{n} X_k\left(\xi_k - \frac{1}{n}\sum_{j=1}^{n} \xi_j\right)\right)\right]E[e^{i\eta\bar{X}}]$$

$$= E\left[\exp\left(i \sum_{k=1}^{n} \xi_k(X_k - \bar{X})\right)\right]E[e^{i\eta\bar{X}}]$$

to verify the assertion.

Mixed moments of a family of random variables

Suppose that $X_1, \ldots, X_n$ are a family of jointly distributed random variables. Their joint characteristic function may be used to find the mixed moment $E[X_1 \cdots X_n]$. As in the case of a single random variable, we differentiate the joint characteristic function and set the ξ_i's $= 0$:

$$\frac{d^n}{d\xi_1 \cdots d\xi_n} C_X(\xi) = E[i^n(X_1 \cdots X_n)e^{i(\xi_1 X_1 + \cdots + \xi_n X_n)}]$$

so that

$$\frac{1}{i^n}\frac{d^n}{d\xi_1 \cdots d\xi_n} C_X(\xi)\bigg|_{\xi_1 = \cdots \xi_n = 0} = E[X_1 \cdots X_n] \qquad (1.8.7)$$

The ξ_i's do not have to be distinct for this formula to hold, as the following example illustrates.

It is not hard to show that the characteristic function of a mean zero bivariate normal random variable is of the form

$$C_{X,Y}(\xi, \eta) = \exp\left[-\tfrac{1}{2}(\alpha\xi^2 + 2\beta\xi\eta + \gamma\eta^2)\right]$$

Assuming that this is the correct form, one can use (1.8.7) to find the values of α, β, and γ. Taking derivatives we find that

$$-\frac{\partial^2}{\partial\xi^2} C_{X,Y}(\xi, \eta)\bigg|_{\xi=0} = \alpha$$

$$-\frac{\partial^2}{\partial\xi\,\partial\eta} C_{X,Y}(\xi, \eta)\bigg|_{\xi=\eta=0} = \beta$$

$$-\frac{\partial^2}{\partial\eta^2} C_{X,Y}(\xi, \eta)\bigg|_{\eta=0} = \gamma$$

and so

$$\alpha = E[X^2]$$
$$\beta = E[XY]$$
$$\gamma = E[Y^2]$$

PROBLEMS

1.1. Show that

$$\int_0^\infty \left|\frac{\sin x}{x}\right| = \infty$$

1.2. Prove that

(a) $\lim_{x\to\infty} \dfrac{1 - \cos x}{x} = 0$

(b) $\lim_{x\to0} \dfrac{1 - \cos x}{x} = 0$

(c) $\lim\limits_{x \to 0} \dfrac{1 - \cos x}{x^2} = \dfrac{1}{2}$

(d) $\dfrac{1 - \cos x}{x^2} \le \dfrac{1}{2},\ x \ne 0$

1.3. Show that the integrability assumption is necessary in Abel's continuity theorem by finding a nonintegrable function f for which

$$\int_0^\infty f(x)\ dx \ne \lim_{\epsilon \to 0+} \int_0^\infty e^{-\epsilon x} f(x)\ dx$$

Hint: Let f be the square-wave function.

1.4. Show that $\displaystyle\int_0^\infty \dfrac{\sin \alpha \omega}{\omega} \sin \omega x\ d\omega$ exists.

1.5. Verify directly, by integration, that the Fourier inversion formula holds for the unit rectangle function

$$f(x) = \begin{cases} 1, & -1 < x < 1 \\ \frac{1}{2}, & x = -1,\ +1 \\ 0, & \text{else} \end{cases}$$

1.6. If $C_X(\xi)$ is the characteristic function of a random variable X with density function f_X, show that

(a) $C_X^{(n)}(\xi) = i^n \displaystyle\int_{-\infty}^\infty e^{i\xi x} x^n f_X(x)\ dx$

(b) $\left| C_X^{(n)}(0) \right| = \displaystyle\int_{-\infty}^\infty x^n f_X(x)\ dx$

1.7. If X has a symmetric density function, show that its characteristic function is real.

1.8. Let $C_X(\xi)$ be the characteristic function of a random variable X. Show that
(a) C_X is uniformly continuous.
(b) $C_X(0) = 1$
(c) $|C_X| \le 1$
(d) C_X is positive definite, that is,

$$\sum_{j=1}^n \sum_{k=1}^n C_X(\xi_j - \xi_k) z_j \bar{z}_k \ge 0$$

for all real $\xi_1, \ldots, \xi_n$ and complex $z_1, \ldots, z_n$.

1.9. If $C_X(\xi) = u(\xi) + iv(\xi)$ show that
(a) $aX + b$ has characteristic function $e^{ib\xi} C_X(a\xi)$.
(b) $\bar{C}_X$ is the characteristic function of $-X$.

(c) u is even and v is odd.

(d) $0 \le 1 + u(2\xi) \le 4[1 + u(\xi)]$

1.10. Suppose that X and Y are independent identically distributed random variables with symmetric distribution of mean zero and variance 1. Let C be their common characteristic function. Suppose that $X + Y$ and $X - Y$ are independent.

(a) Show that $Z_1 = X + Y$, $Z_2 = X - Y$ each have characteristic function $C_Z(\xi) = C^2(\xi)$.

(b) Show that $W = (X + Y) + (X - Y)$ has characteristic function $C_W(\xi) = C^4(\xi) = C(2\xi)$.

(c) Solve the equation in part (b) and conclude that X and Y are normally distributed.

1.11. Evaluate the characteristic function of a normal random variable of mean zero and variance 1 by complex integration.

1.12. Perform a change of variables on the formula for the characteristic function of a normal random variable of mean zero and variance 1 to find the characteristic function of a normal random variable of mean μ and variance σ^2.

1.13. Find the characteristic function of a Poisson random variable.

1.14. If X is a discrete random variable with

$$P\left[X = \frac{k}{n}\right] = \frac{1}{n}, \qquad k = 1, \ldots, n$$

find its characteristic function, and show that as $n \to \infty$ it converges to the characteristic function of a random variable uniform on $[0, 1]$.

1.15. A random variable X is said to have a Cauchy distribution if its probability density function is

$$f_X(x) = \frac{k}{\pi[k^2 + (x - \mu)^2]}$$

Show that the characteristic function for this random variable is

$$C_X(\xi) = e^{i\mu\xi - k|\xi|}$$

1.16. Suppose that X has the geometric distribution

$$P[X = k] = (1 - p)^{k-1}p$$

If $p = \lambda/n$, show that X/n becomes exponential as $n \to \infty$ by using the characteristic function.

1.17. Use the characteristic function to show that in the limit a binomial random variable $B_{n,\lambda/n}$ becomes Poisson.

1.18. Use the characteristic function to show that if X_1, X_2, . . . are independent identically distributed mean zero normal random variables, then

$$\frac{X_1 + \cdots + X_n}{\sqrt{n}}$$

has the same distribution as any of the X_i's.

1.19. Suppose that X and Y are independent normal random variables. Use the characteristic function to find the distributions of

(a) X/Y.

(b) X^{-2}.

1.20. If $F(\omega)$ is the Fourier transform of $f(x)$, show that if the various transforms exist, then

(a) The Fourier transform of $e^{\pm i\omega_0 x} f(x)$ is $F(\omega \pm \omega_0)$.

(b) The Fourier transform of $f'(x)$ is $-i\omega F(\omega)$.

(c) The Fourier transform of $\int_{-\infty}^{x} f(t)\, dt$ is $F(\omega)/-i\omega$.

1.21. Use the results of Problem 1.20 to show another form of the inversion formula for characteristic functions

$$\frac{F(x + h) - F(x)}{h} = \frac{1}{2\pi} \int_{-\infty}^{\infty} C_X(\xi) \frac{1 - e^{i\xi h}}{i\xi h} e^{-i\xi x}\, d\xi$$

where F is the distribution function of the random variable X.

1.22. Use the characteristic function to find the density for the sum of n independent identically distributed

(a) exponential random variables.

(b) normal random variables.

(c) Cauchy random variables.

(d) Poisson random variables.

1.23. Use the continuity theorem for characteristic functions to show that as $\lambda \to \infty$,

$$W = \frac{N_\lambda - \lambda}{\sqrt{\lambda}}$$

becomes normal of mean zero and variance 1.

1.24. If $Z = \sum_{k=1}^{n} a_k X_k$ and the X_k's are independent, express the characteristic function of Z in terms of the characteristic function of the X_k's.

1.25. If the characteristic function of a random variable X is of the form

$$C_X(\xi) = \frac{\sin n\xi}{n\xi}$$

show that X is uniform on $[-n, n]$.

1.26. If X_k are independent, with characteristic functions $C_{X_k}(\xi)$, respectively, describe the random variables that have the characteristic functions:

(a) $\displaystyle\prod_{k=1}^{n} C_{X_k}$.

(b) $|C_{X_k}|^2$.

(c) $\displaystyle\sum_{k=1}^{n} p_k C_{X_k}$ if $\displaystyle\sum_{k=1}^{n} p_k = 1$.

1.27. Given the Fourier inversion formula holds for $f(x)$, show that

$$f(x) = \frac{1}{\pi} \int_0^\infty (A(\xi) \cos \xi x + B(\xi) \sin \xi x) \, d\xi$$

where

$$A(\xi) = \int_{-\infty}^{\infty} f(y) \cos \xi y \, dy, \qquad B(\xi) = \int_{-\infty}^{\infty} f(y) \sin \xi y \, dy$$

This is called the Fourier integral representation of $f(x)$.

1.28. Find the Fourier integral representation of $f(x) = e^{-k|x|}$, $k > 0$.

CHAPTER

2

Fourier Series

One of the most important problems in analysis is to find a way to approximate "arbitrary" functions by linear combinations of some prescribed functions. Taylor's series is a familiar example of this, where the approximating functions are linear combinations of the polynomials $(x - a)^k$, and is suitable for functions which have many derivatives. Fourier series approximates by linear combinations of $\sin kt$ and $\cos kt$, and turns out to be useful for approximating a large class of functions, including those which are discontinuous at a number of points. Further, the study of these series will lead us to the Fourier inversion formula in a totally different way than what we have seen in Chapter 1.

2.1 THE MEAN SQUARE APPROXIMATION

Let $\phi_0(t)$, $\phi_1(t)$, . . . be real-valued functions defined on an interval $[a, b]$. These functions are said to be orthonormal (orthogonal and normal) on $[a, b]$ if

$$\int_a^b \phi_i(t)\phi_j(t)\, dt = \begin{cases} 1, & i = j \\ 0, & i \neq j \end{cases} \tag{2.1.1}$$

A real-valued function f is said to be square integrable on $[a, b]$ if

$$\int_a^b [f(t)]^2\, dt < \infty \tag{2.1.2}$$

In this section we consider the general problem of approximating a given square-integrable function f by linear combinations

$$\sum_{k=0}^{n} a_k \phi_k(t)$$

of a collection of orthonormal functions $\{\phi_0, \phi_1, \ldots\}$. We shall take

$$\int_a^b \left[f(t) - \sum_{k=0}^{n} a_k \phi_k(t) \right]^2 dt \qquad (2.1.3)$$

to be a measure of the accuracy of the approximation and call a particular linear combination

$$\sum_{k=0}^{n} c_k \phi_k(t)$$

the mean square approximation to f by $\{\phi_0, \ldots, \phi_n\}$ if

$$\int_a^b \left[f(t) - \sum_{k=0}^{n} c_k \phi_k(t) \right]^2 dt \leq \int_a^b \left[f(t) - \sum_{k=0}^{n} a_k \phi_k(t) \right]^2 dt$$

for any choice of $a_0, a_1, \ldots, a_n$. We can find the values of the c_k's for this optimal approximation as follows. Write (2.1.3) as

$$\int_a^b \left[f(t) - \sum_{k=0}^{n} a_k \phi_k(t) \right]^2 dt$$

$$= \int_a^b f^2(t)\, dt - 2 \sum_{k=0}^{n} a_k \int_a^b f(t)\phi_k(t)\, dt$$

$$+ \int_a^b \left[\sum_{k=0}^{n} a_k \phi_k(t) \right]^2 dt$$

which since the ϕ_k's are orthonormal is

$$= \int_a^b f^2(t)\, dt - 2 \sum_{k=0}^{n} a_k \int_a^b f(t)\phi_k(t)\, dt + \sum_{k=0}^{n} a_k^2$$

and completing the square for each a_k yields

$$= \int_a^b f^2(t) \, dt + \sum_{k=0}^n \left[a_k - \int_a^b f(t)\phi_k(t) \, dt \right]^2$$

$$- \sum_{k=0}^n \left[\int_a^b f(t)\phi_k(t) \, dt \right]^2$$

Choosing the a_k's to make the middle term zero minimizes (2.1.3); thus

$$c_k = \int_a^b f(t)\phi_k(t) \, dt \qquad (2.1.4)$$

If we have an infinite orthonormal system $\{\phi_0, \phi_1, \ldots\}$, it may happen that with this choice of c_k's, $k = 0, 1, 2, \ldots$,

$$\lim_{n \to \infty} \int_a^b \left[f(t) - \sum_{k=0}^n c_k\phi_k(t) \right]^2 dt = 0 \qquad (2.1.5)$$

In this case we say that $f(t)$ is exactly represented by

$$\sum_{k=0}^\infty c_k\phi_k(t) \qquad (2.1.6)$$

in the mean square sense, even though this does not imply that this series converges for any specified t. We shall call (2.1.6) the mean square approximation to f by the orthonormal system $\{\phi_0, \phi_1, \ldots\}$. If equation (2.1.5) is true for any square-integrable function f, the orthonormal system is called complete.

2.2   FOURIER SERIES

Since

$$\int_0^{2\pi} \cos mt \cos kt \, dt = \begin{cases} 0, & m \neq k \\ \pi, & m = k = 1, 2, 3, \ldots \\ 2\pi, & m = k = 0 \end{cases}$$

$$\int_0^{2\pi} \sin mt \sin kt \, dt = \begin{cases} 0, & m \neq k \\ \pi, & m = k = 1, 2, 3, \ldots \\ 0, & m = k = 0 \end{cases}$$

$$\int_0^{2\pi} \sin kt \cos mt \, dt = 0, \qquad \text{all } m \text{ and } k$$

it follows that the functions

$$\frac{1}{\sqrt{2\pi}}, \quad \frac{\cos t}{\sqrt{\pi}}, \quad \frac{\sin t}{\sqrt{\pi}}, \quad \frac{\cos 2t}{\sqrt{\pi}}, \quad \frac{\sin 2t}{\sqrt{\pi}}, \dots \tag{2.2.1}$$

form an orthonormal set on $[0, 2\pi]$. Suppose that f is defined on $[0, 2\pi]$ and is square integrable. According to Section 2.1, the mean square approximation to f by this orthonormal set is

$$\frac{a_0'}{\sqrt{2\pi}} + \sum_{k=1}^{\infty} a_k' \frac{\cos kt}{\sqrt{\pi}} + b_k' \frac{\sin kt}{\sqrt{\pi}} \tag{2.2.2}$$

where, by (2.1.4),

$$a_0' = \int_0^{2\pi} f(t) \frac{1}{\sqrt{2\pi}} dt$$

$$a_k' = \int_0^{2\pi} f(t) \frac{\cos kt}{\sqrt{\pi}} dt \tag{2.2.3}$$

$$b_k' = \int_0^{2\pi} f(t) \frac{\sin kt}{\sqrt{\pi}} dt$$

In order to obtain a simpler form and to follow standard notation, we rewrite (2.2.2) as

$$\frac{a_0}{2} + \sum_{k=1}^{\infty} a_k \cos kt + b_k \sin kt \tag{2.2.4}$$

where the coefficients are now given by the formulas

$$a_0 = \frac{1}{\pi} \int_0^{2\pi} f(t) \, dt$$

$$a_k = \frac{1}{\pi} \int_0^{2\pi} f(t) \cos kt \, dt \tag{2.2.5}$$

$$b_k = \frac{1}{\pi} \int_0^{2\pi} f(t) \sin kt \, dt$$

The infinite series (2.2.4) is called the Fourier series of $f(t)$, its coefficients are called the Fourier coefficients of $f(t)$, and formulas (2.2.5) are called the Euler formulas for these coefficients.

2.3 CONVERGENCE OF FOURIER SERIES, CONSEQUENCES

Perhaps the most important property of the set of functions (2.2.4) is that they are complete. We shall not prove this fact. Recall completeness means that any square-integrable function $f(t)$ defined on $[0, 2\pi]$ can be exactly represented by a Fourier series, in the mean square sense:

$$0 = \lim_{n \to \infty} \int_0^{2\pi} \left[f(t) - \left(\frac{a_0}{2} + \sum_{k=1}^n a_k \cos kt + b_k \sin kt \right) \right]^2 dt$$

$$= \lim_{n \to \infty} \left[\int_0^{2\pi} f^2(t) \, dt - \frac{\pi a_0^2}{2} - \sum_{k=1}^n \pi a_k^2 - \pi b_k^2 \right] \tag{2.3.1}$$

or

$$\frac{1}{\pi} \int_0^{2\pi} f^2(t) \, dt = \frac{a_0^2}{2} + \sum_{k=1}^\infty a_k^2 + b_k^2 \tag{2.3.2}$$

which is known as **Parseval's identity.**

If we let f_n denote the partial sum

$$f_n(t) = \frac{a_0}{2} + \sum_{k=1}^n a_k \cos kt + b_k \sin kt$$

then (2.3.1) becomes

$$\lim_{n \to \infty} \int_0^{2\pi} [f(t) - f_n(t)]^2 = 0 \tag{2.3.3}$$

and we say that f_n converges to f in the mean square. What we really would like, however, is

$$\lim_{n \to \infty} f_n(t) = f(t) \tag{2.3.4}$$

to hold for each t, which is called pointwise convergence. Unfortunately, pointwise convergence and mean square convergence are not closely related, so we cannot conclude one from the other based on general principles. The following theorem gives conditions which are sufficient for the pointwise convergence of Fourier series, and we state it without proof.

If $f(t)$ is periodic with period 2π, and "smooth" (e.g., f and f' are bounded continuous for all t, except for a finite number of points in $[0, 2\pi]$), then

$$\frac{a_0}{2} + \sum_{k=1}^\infty a_k \cos kt + b_k \sin kt = \frac{f(t-) + f(t+)}{2} \tag{2.3.5}$$

where $f(t-) = \lim_{s \to t-} f(s)$ and $f(t+) = \lim_{s \to t+} f(s)$. Thus the Fourier series converges to $f(t)$ when t is a continuity point of $f(t)$. As an illustration of this theorem we present the following example.

The square-wave function and Euler's formula

Consider the ''square-wave'' function

$$f(t) = \begin{cases} 1, & 2n\pi \le t < (2n+1)\pi \\ -1, & (2n+1)\pi \le t < (2n+2)\pi, \end{cases} \quad n = \ldots, -2, -1, 0, 1, 2, \ldots$$

(2.3.6)

which is periodic with period 2π (see Fig. 2.1). By (2.2.5) the Fourier series for f has $a_k = 0$ for all k, and

$$b_k = \frac{2 - 2\cos k\pi}{k\pi} = \begin{cases} 0, & k \text{ even} \\ \dfrac{4}{k\pi}, & k \text{ odd} \end{cases}$$

Applying Parseval's identity (2.3.2) to the square-wave function, we get

$$\frac{1}{\pi} \int_0^{2\pi} f^2(t)\, dt = \frac{16}{\pi^2}\left(1 + \frac{1}{3^2} + \frac{1}{5^2} + \cdots\right)$$

and since

$$\frac{1}{\pi} \int_0^{2\pi} f^2(t)\, dt = 2$$

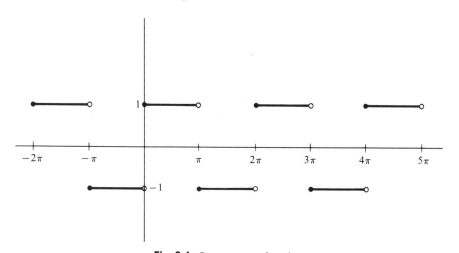

Fig. 2.1. Square-wave function

this becomes

$$\sum_{k=1}^{\infty} \frac{1}{(2k-1)^2} = \frac{\pi^2}{8}$$

Applying the identity

$$\frac{3}{4} \sum_{k=1}^{\infty} \frac{1}{k^2} = \sum_{k=1}^{\infty} \frac{1}{(2k-1)^2}$$

obtained by considering

$$\sum_{k=1}^{\infty} \frac{1}{k^2}$$

to be a sum over even and odd k's separately, we get

$$\sum_{k=1}^{\infty} \frac{1}{k^2} = \frac{\pi^2}{6} \tag{2.3.7}$$

which is known as **Euler's formula**.

Since f satisfies the hypothesis of the theorem, the series will converge to f at each t where f is continuous, so for such t

$$f(t) = \sum_{k=\text{odd}}^{\infty} \frac{4}{k\pi} \sin kt = \frac{4}{\pi} (\sin t + \tfrac{1}{3} \sin 3t + \cdots)$$

For $t = \pi/2$ this becomes

$$1 = \sum_{k=\text{odd}}^{\infty} \frac{4}{k\pi} \sin k\left(\frac{\pi}{2}\right) = \frac{4}{\pi} (1 - \tfrac{1}{3} + \tfrac{1}{5} - \tfrac{1}{7} + \cdots) \tag{2.3.8}$$

We can also verify this directly, by comparing

$$\int_0^1 \frac{dx}{1 + x^2} = \tan^{-1}x \Big|_0^1 = \frac{\pi}{4}$$

and

$$\int_0^1 \frac{dx}{1 + x^2} = \int_0^1 (1 - x^2 + x^4 - x^6 + \cdots)\, dx$$
$$= 1 - \tfrac{1}{3} + \tfrac{1}{5} - \tfrac{1}{7} + \cdots$$

At points where f is discontinuous, say, for example, $t = \pi$, according to the theorem the Fourier series sums to the average of the left- and right-hand limits, that is, we should have

$$\frac{4}{\pi} (\sin \pi + \tfrac{1}{3} \sin 3\pi + \cdots) = \frac{f(\pi-) + f(\pi+)}{2}$$

which we may also verify directly, obtaining zero on each side.

2.4 FUNCTIONS OF ARBITRARY PERIOD

We now derive the Fourier series for functions of arbitrary period. Suppose that f is periodic with period T, continuous, and "smooth." The function

$$g(t) = f\left(\frac{tT}{2\pi}\right)$$

then has period 2π and is "smooth," so the theorem applies and

$$g(t) = \frac{a_0}{2} + \sum_{k=1}^{\infty} a_k \cos kt + b_k \sin kt \qquad (2.4.1)$$

at continuity points of g, where

$$a_0 = \frac{1}{\pi} \int_0^{2\pi} g(t)\, dt$$

$$a_k = \frac{1}{\pi} \int_0^{2\pi} g(t) \cos kt\, dt \qquad (2.4.2)$$

$$b_k = \frac{1}{\pi} \int_0^{2\pi} g(t) \sin kt\, dt$$

We may rewrite this, by the change of scale $s = tT/2\pi$, as

$$f(s) = \frac{a_0}{2} + \sum_{k=1}^{\infty} a_k \cos \frac{2\pi ks}{T} + b_k \sin \frac{2\pi ks}{T} \qquad (2.4.3)$$

and put the integrals (2.4.2) in the form

$$a_0 = \frac{2}{T} \int_0^T f(s)\, ds$$

$$a_k = \frac{2}{T} \int_0^T f(s) \cos \frac{2\pi ks}{T}\, ds \qquad (2.4.4)$$

$$b_k = \frac{2}{T} \int_0^T f(s) \sin \frac{2\pi ks}{T}\, ds$$

The series in (2.4.3) is called the Fourier series for a function f with period T, and we note that this series converges to the average of the left- and right-hand limits, if the function f is smooth.

2.5   FOURIER SERIES FOR ARBITRARY FUNCTIONS

If f is not periodic, but smooth, then a consequence of the convergence theorem is that

$$f(s) = \frac{a_0}{2} + \sum_{k=1}^{n} a_k \cos \frac{2\pi ks}{T} + b_k \sin \frac{2\pi ks}{T}$$

for any $0 < s < T$ which is a continuity point of f. This is because the function formed by duplicating the values of f in $(0, T]$, outside that interval in a periodic fashion, satisfies the hypothesis of the theorem. Of course, this is not the only periodic function that agrees with f on $(0, T]$. For example, one can consider the even periodic extension, which has period $2T$,

$$g(x) = \begin{cases} f(x), & x \in (0, T] \\ f(-x), & x \in (-T, 0] \end{cases}$$

or the odd periodic extension, which also has period $2T$,

$$h(x) = \begin{cases} f(x) & x \in (0, T] \\ -f(-x), & x \in (-T, 0] \end{cases}$$

These can be seen in Figs. 2.2 through 2.4, where for illustrative purpose we have taken $f(x) = x + 1$ and the interval $(0, 1]$. Corresponding to each of these is a different Fourier series, and thus there are many possible Fourier series representing a given function on the interval $(0, T)$. We shall extend our functions only by duplication.

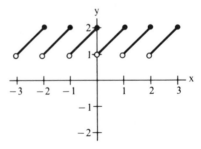

Fig. 2.2 Extension by duplication

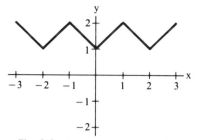

Fig. 2.3. Even periodic extension

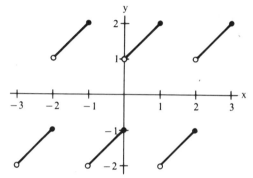

Fig. 2.4. Odd periodic extension

2.6   THE COMPLEX FORM OF THE FOURIER SERIES

The Fourier series for $f(t)$ takes on a simple form when expressed in terms of

$$e^{ix} = \cos x + i \sin x$$

It is easy to verify that

$$\sum_{k=-\infty}^{\infty} \alpha_k e^{-ikt} \tag{2.6.1}$$

is the same as (2.2.4), where

$$\alpha_k = \frac{1}{2\pi} \int_0^{2\pi} f(t) e^{ikt} \, dt \tag{2.6.2}$$

To show this, we note that $\alpha_0 = a_0/2$ and

$$\alpha_k = \tfrac{1}{2}(a_k - ib_k)$$
$$\alpha_{-k} = \tfrac{1}{2}(a_k + ib_k), \qquad k = 1, 2, \ldots$$

so that

$$\alpha_k e^{ikt} = \tfrac{1}{2}[a_k \cos kt + b_k \sin kt + i(-b_k \cos kt + a_k \sin kt)]$$
$$\alpha_{-k} e^{-ikt} = \tfrac{1}{2}[a_k \cos kt + b_k \sin kt + i(b_k \cos kt - a_k \sin kt)]$$

implying that

$$\sum_{k=-\infty}^{\infty} \alpha_k e^{-ikt} = \frac{a_0}{2} + \sum_{k=1}^{\infty} a_k \cos kt + b_k \sin kt$$

For functions of period T we express the corresponding formulas as

$$\sum_{k=-\infty}^{\infty} \alpha(\omega_k) e^{-i\omega_k t}, \qquad \omega_k = \frac{2\pi k}{T} \tag{2.6.3}$$

where

$$\alpha(\omega_k) = \frac{1}{T} \int_0^T f(t) e^{i\omega_k t}\, dt \tag{2.6.4}$$

2.7   THE FOURIER INVERSION FORMULA

Since f is assumed periodic with period T, (2.6.4) can be written as

$$\alpha(\omega_k) = \frac{1}{T} \int_{-T/2}^{T/2} f(t) e^{i\omega_k t}\, dt \tag{2.7.1}$$

Given a smooth continuous function f defined on $(-T/2, T/2]$ we may extend it, by duplication, to the whole real line to have period T. The theorem then applies and we can write, for $x \in (-T/2, T/2]$,

$$f(x) = \sum_{k=-\infty}^{\infty} \alpha(\omega_k) e^{-i\omega_k x} \tag{2.7.2}$$

which by (2.7.1)

$$= \sum_{k=-\infty}^{\infty} \left[\frac{1}{T} \int_{-T/2}^{T/2} f(t) e^{i\omega_k t}\, dt \right] e^{-i\omega_k x}$$

or

$$= \frac{1}{2\pi} \sum_{k=-\infty}^{\infty} \left[\int_{-T/2}^{T/2} f(t) e^{-i\omega_k t}\, dt \right] e^{-i\omega_k x} \frac{2\pi}{T}$$

Now, defining $\tilde{F}$ to be the approximate Fourier transform of f given by

$$\tilde{F}(\omega_k) = \int_{-T/2}^{T/2} f(t)e^{i\omega_k t}\,dt \qquad (2.7.3)$$

we see that (2.7.2) is a Riemann sum

$$f(x) = \frac{1}{2\pi} \sum_{k=-\infty}^{\infty} \tilde{F}(\omega_k)e^{-i\omega_k x}\,\Delta\omega$$

where $\Delta\omega = 2\pi/T$. As $T \to \infty$, assuming that $\tilde{F} \to F$ sufficiently well, this equation becomes

$$f(x) = \frac{1}{2\pi} \int_{-\infty}^{\infty} F(\omega)e^{-i\omega x}\,d\omega$$

$$= \frac{1}{2\pi} \int_{-\infty}^{\infty} \left[\int_{-\infty}^{\infty} f(t)e^{i\omega t}\,dt \right] e^{-i\omega x}\,d\omega$$

which is the Fourier inversion formula.

2.8   PLANCHEREL'S FORMULA

A consequence of completeness is that the square of f and the square of its Fourier series have the same integral:

$$\int_{0}^{T} f^2(t)\,dt = \int_{0}^{T} \left[\sum_{k=-\infty}^{\infty} \alpha(\omega_k)e^{-i\omega_k t} \right]^2 dt$$

$$= T \sum_{k=-\infty}^{\infty} \alpha(\omega_k)\alpha(\omega_{-k})$$

noting that $\alpha(\omega_{-k}) = \overline{\alpha(\omega_k)}$

$$= T \sum_{k=-\infty}^{\infty} |\alpha(\omega_k)|^2$$

so

$$\frac{1}{T} \int_{0}^{T} f^2(t)\,dt = \sum_{k=-\infty}^{\infty} |\alpha(\omega_k)|^2 \qquad (2.8.1)$$

which is the complex form of Parseval's identity. Substituting the value of $\alpha(\omega_k)$ from (2.6.4) into this, we obtain

$$\frac{1}{T} \int_0^T f^2(t) \, dt = \sum_{k=-\infty}^{\infty} \left| \frac{1}{T} \int_0^T f(t) e^{i\omega_k t} \, dt \right|^2$$

or since f is periodic,

$$\frac{1}{T} \int_{-T/2}^{T/2} f^2(t) \, dt = \sum_{k=-\infty}^{\infty} \left| \frac{1}{T} \int_{-T/2}^{T/2} f(t) e^{-i\omega_k t} \, dt \right|^2$$

and using the notation (2.7.3)

$$\int_{-T/2}^{T/2} f^2(t) \, dt = \frac{1}{2\pi} \sum_{k=-\infty}^{\infty} |\tilde{F}(\omega_k)|^2 \frac{2\pi}{T}$$

Recognizing the right-hand side as a Riemann sum, and assuming that $\tilde{F}$ converges to F sufficiently well, when $T \to \infty$ we obtain **Plancherel's formula:**

$$\int_{-\infty}^{\infty} f^2(t) \, dt = \frac{1}{2\pi} \int_{-\infty}^{\infty} |F(\omega)|^2 \, d\omega$$

$$= \frac{1}{2\pi} \int_{-\infty}^{\infty} \left| \int_{-\infty}^{\infty} f(t) e^{i\omega t} \, dt \right|^2 d\omega \qquad (2.8.2)$$

which can be regarded as a continuous version of Parseval's identity (2.3.2).

Using this and its obvious counterpart,

$$\int_{-\infty}^{\infty} F^2(t) \, dt = 2\pi \int_{-\infty}^{\infty} \left| \frac{1}{2\pi} \int_{-\infty}^{\infty} F(t) e^{-i\omega t} \, dt \right|^2 d\omega$$

we can verify the statement, given in Section 1.3, that we may use the inversion formula with square-integrable probability density functions. Letting

$$h(x) = \frac{1}{2\pi} \int_{-\infty}^{\infty} \int_{-\infty}^{\infty} f(y) e^{i\omega y} \, dy \, e^{-i\omega x} \, d\omega \qquad (2.8.3)$$

be the result of the Fourier inversion formula, f_k a sequence of step functions approximating f in the mean square, and F_k their Fourier transforms,

$$\int_{-\infty}^{\infty} [f(t) - f_k(t)]^2 \, dt = \frac{1}{2\pi} \int_{-\infty}^{\infty} |F(\omega) - F_k(\omega)|^2 \, d\omega \qquad (2.8.4)$$

which by (2.8.3)

$$= \int_{-\infty}^{\infty} |h(t) - f_k(t)|^2 \, dt$$

Using the "triangle" inequality,

$$\left[\int_{-\infty}^{\infty} (p(t) + q(t))^2 \, dt \right]^{1/2} \leq \left[\int_{-\infty}^{\infty} p^2(t) \, dt \right]^{1/2} + \left[\int_{-\infty}^{\infty} q^2(t) \, dt \right]^{1/2} \quad (2.8.5)$$

which is valid for any functions p and q,

$$0 \leq \left[\int_{-\infty}^{\infty} |f(t) - h(t)|^2 \, dt \right]^{1/2}$$

$$\leq \left[\int_{-\infty}^{\infty} |f(t) - f_k(t)|^2 \, dt \right]^{1/2} + \left[\int_{-\infty}^{\infty} |h(t) - f_k(t)|^2 \, dt \right]^{1/2}$$

and by (2.8.4)

$$= 2 \left[\int_{-\infty}^{\infty} |f(t) - f_k(t)|^2 \, dt \right]^{1/2}$$

Since

$$\lim_{k \to \infty} \int_{-\infty}^{\infty} |f(t) - f_k(t)|^2 \, dt = 0$$

we have

$$\int_{-\infty}^{\infty} |f(t) - h(t)|^2 \, dt = 0$$

which is what was required.

2.9 THE GEOMETRY OF FOURIER SERIES, HILBERT SPACES

A vector $(x_1, \ldots, x_n)$ in $\mathbb{R}^n$ can be thought of as a function $x(k)$ from $\{1, \ldots, n\}$ into $\mathbb{R}$. With this interpretation, the usual operations of addition and scalar multiplication in $\mathbb{R}^n$ are the operations of addition and scalar multiplication of functions.

The space of such functions is called finite dimensional since every function $x(k)$ can be represented as a linear combination of a fixed finite number of "coordinate functions" e_i

$$x(k) = \sum_{i=1}^{n} b_i e_i(k)$$

where

$$e_i(k) = \begin{cases} 1, & k = i \\ 0, & k \neq i \end{cases}$$

The functions e_i are said to span the space of functions $\{x(k)\}$. In fact, they are a basis, by which we mean the e_i's are also linearly independent as functions of k:

$$\sum_{i=1}^{n} b_i e_i(k) = 0 \quad \text{if and only if} \quad b_i = 0 \qquad \text{for all } i$$

This implies that the b_i's appearing in the representation of $x(k)$ are unique.

It is clear that all the familiar properties of $\mathbb{R}^n$ can be expressed in terms of $x(k)$; in particular the inner product is given by

$$<x(k),\, y(k)> = \sum_{k=1}^{n} x(k)y(k)$$

and the length $\|x(k)\|$ by

$$\|x(k)\| = \sqrt{\sum_{k=1}^{n} [x(k)]^2}$$

Recall that two vectors are called orthogonal if their inner product is zero. The set $\{e_i(k)\}$ is called an orthonormal basis, which means that in addition to being a basis, it satisfies

$$<e_i(k),\, e_j(k)> = \begin{cases} 1, & i = j \\ 0, & i \neq j \end{cases}$$

This implies that

$$\|x(k)\| = \sqrt{\sum_{k=1}^{n} b_k^2}$$

Another important property of orthonormal bases is that the b_i's can be found from the length of the projection of x in the e_i direction:

$$b_i = <x(k), e_i(k)>$$

The question arises whether there are other spaces of functions for which one can define notions of inner product and length, and still have the familiar geometric properties of these quantities preserved. The answer is yes, and such spaces of functions are called Hilbert spaces.

The essential difficulty in generalizing what we have seen above to other spaces of functions is that function spaces are usually infinite dimensional. This means that the basis has an infinite number of elements, and that the sums involved will have an infinite number of terms. The situation becomes technically more difficult by the limits that must be considered.

The abstract definition is as follows. A (real) Hilbert space is a (real) vector space X, such that

(i)   There is an inner product denoted $<x, y>$ which is real valued, and satisfies

$$<x, y> = <y, x>$$
$$<ax + by, z> = a<x, z> + b<y, z>$$
$$<x, x> > 0 \qquad \text{iff } x \neq 0$$

(ii)   The norm defined by

$$\|x\| = <x, x>^{1/2}$$

makes H complete, meaning that by using the metric

$$d(x, y) = \|x - y\|$$

H becomes a complete metric space.

We make some comments on the definition. Saying that H is a vector space means that addition, subtraction, and scalar multiplication are defined on H, with the obvious properties. In particular there is an additive identity, denoted $\mathbf{0}$. Saying that $\|\cdot\|$ is a norm means that it is positive $\|x\| \geq 0$, satisfies the triangle inequality

$$\|x + y\| \leq \|x\| + \|y\|$$

and that $\|x\| = 0$ implies that $x = \mathbf{0}$. Calling H complete means that Cauchy sequences converge, that is,

$$\|x_n - x_m\| \to 0 \qquad \text{as } n, m \to \infty$$

implies that there exists $x \in H$ such that

$$\|x - x_n\| \to 0 \qquad \text{as } n \to \infty$$

A subspace of H that is closed with respect to Cauchy sequences is called a closed subspace. We state without proof the important **projection theorem,** which says if M is a closed subspace of H that any element x of H can be uniquely represented as $x = y + z$, where y is in M and z is orthogonal to M.

We can put our investigation of Fourier series into this general framework. Consider the space H of square-integrable real-valued functions defined on $[0, 2\pi]$. Define the inner product of two functions f and g in H by

$$<f(x), g(x)> = \int_0^{2\pi} f(x)g(x)\, dx$$

From our discussion, with this inner product H becomes a Hilbert space, completeness of H following from completeness of the orthonormal system (2.2.1). In fact, (2.2.1) is an orthonormal basis for H, and the Fourier coefficients (2.2.3) are the lengths of the projections onto the elements of this basis.

If H is a complex vector space, the inner product satisfies

$$<x, y> = \overline{<y, x>}$$

$$<ax + by, z> = a<x, z> + b<y, z>$$

$$<x, \bar{x}> \text{ is real and nonnegative}$$

$$<x, \bar{x}> \neq 0 \qquad \text{iff } x \neq 0$$

where a and b are complex numbers and the bar denotes complex conjugation. The norm is defined by

$$\|x\| = <x, \bar{x}>^{1/2}$$

The corresponding definition for square-integrable complex-valued functions is

$$<f, g> = \int_0^{2\pi} f(x)\overline{g(x)}\, dx$$

These have the orthonormal basis

$$e_k = \frac{1}{\sqrt{2\pi}} e^{-ikx}, \qquad k = \ldots, -2, -1, 0, 1, 2, \ldots$$

and the projections onto this basis give the form of the Fourier series (2.6.1).

2.10   POSITIVE DEFINITE FUNCTIONS

A complex-valued function f on the real numbers is called positive definite if

$$\sum_{j=1}^{n} \sum_{m=1}^{n} f(t_j - t_m) z_j \overline{z_m} \geq 0$$

The most basic positive definite function is

$$f(x) = e^{ix}$$

This is easily verified by the method usually employed to show positive definiteness: The double sum above becomes

$$\left| \sum_{k=1}^{n} e^{it_k} z_k \right|^2$$

which is certainly nonnegative. Similarly, all the functions

$$f_k(x) = e^{ikx}, \qquad -\infty < k < \infty$$

are positive definite.

Positive linear combinations of positive definite functions are positive definite, so we know that if f is represented by a Fourier series

$$f(x) = \sum_{k=-\infty}^{\infty} a_k e^{-ikx}$$

and the a_k are all nonnegative, f is positive definite. This turns out to characterize positive definite periodic functions. Substituting the Fourier series into the definition gives

$$\sum_{k,m,j} a_k e^{ik(t_j - t_m)} z_j \overline{z_m} \geq 0$$

and by setting

$$z_m = e^{-imt_m}$$

we get

$$\sum_{j,k,m} a_k e^{it_j(k-j) - it_m(k-m)} \geq 0$$

By integrating t_j and t_m over $(0, 2\pi)$, nearly all the terms in the sum become 0 and we only get a contribution for each k when $j = k$ and $m = k$, so

$$\sum_{k=1}^{n} a_k \geq 0$$

Setting $z_m^- = e^{imt_m}$ shows that it is true for negative k's as well. This implies that $a_k \geq 0$ for each k. This is a sufficient condition as well. Thus the continuous non-negative definite periodic functions are those with nonnegative Fourier coefficients. This is a version of Bochner's theorem.

PROBLEMS

2.1. Let $f(t)$ be as shown on the graph. Find the Fourier series for $f(t)$ and verify that its sum is equal to $f(t)$ for $t = 0$ and $t = \pi$.

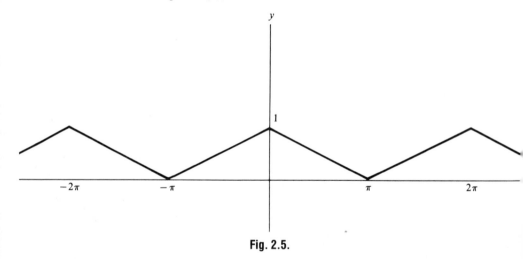

Fig. 2.5.

2.2. Verify Formula (2.6.4).

2.3. Find the Fourier series for the even, odd, and periodic extensions of

$$f(x) = x + 1$$

defined on $0 \leq x \leq 1$.

2.4. Show that if α is not an integer and $0 < x < \pi$, then the Fourier series for $\cos \alpha x$ is

$$\frac{2\alpha \sin \pi\alpha}{\pi} \left[\frac{1}{2\alpha^2} + \sum_{k=1}^{\infty} \frac{(-1)^k \cos kx}{\alpha^2 - k^2} \right]$$

2.5. Obtain the Fourier series for the following functions defined on $-\pi < x < \pi$:

(a) $x^2 \operatorname{sgn} x$.

(b) $\sin x \operatorname{sgn} x$.

(c) e^x.

2.6. Suppose that f and g are defined by

$$f(x) = \begin{cases} 0, & -\pi \le x < 0 \\ 1, & 0 \le x < \pi \end{cases}$$

$$g(x) = \begin{cases} 0, & -\pi \le x < 0 \\ x, & 0 \le x < \pi \end{cases}$$

and have

$$\frac{1}{2} + \frac{2}{\pi} \sum_{n=1}^{\infty} \frac{\sin (2n-1)x}{2n-1}$$

$$\frac{\pi}{4} - \frac{2}{\pi} \sum_{n=1}^{\infty} \frac{\cos (2n-1)x}{(2n-1)^2} - \sum_{n=1}^{\infty} (-1)^n \frac{\sin nx}{n}$$

as their Fourier series, respectively. Without any integration find the Fourier series for the following functions:

$$h(x) = \begin{cases} 1, & -\pi \le x < 0 \\ 0, & 0 \le x < \pi \end{cases}$$

$$k(x) = \begin{cases} a + bx, & -\pi \le x < 0 \\ c + dx, & 0 \le x < \pi \end{cases}$$

2.7. Let g and g' be continuous functions on an interval $[a, b]$. Prove that

$$\lim_{n \to \infty} \int_a^b g(x) \cos nx \, dx = \lim_{n \to \infty} \int_a^b g(x) \sin nx \, dx = 0$$

2.8. Using the Fourier series for

$$f(x) = \begin{cases} 0, & -\pi \le x < 0 \\ x^2, & 0 \le x < \pi \end{cases}$$

deduce the following results:

(a) $\displaystyle\sum_{k=1}^{\infty} \frac{1}{k^2} = \frac{\pi^2}{6}$

(b) $\displaystyle\sum_{k=1}^{\infty} \frac{(-1)^{k+1}}{k^2} = \frac{\pi^2}{12}$

(c) $\displaystyle\sum_{k=1}^{\infty} \frac{1}{(2k-1)^2} = \frac{\pi^2}{8}$

2.9. By applying Parseval's identity, prove for square-integrable functions f and g with Fourier coefficients a_k, b_k and c_k, d_k respectively,

$$\frac{1}{\pi} \int_0^{2\pi} f(t)g(t)\, dt = \frac{a_0 c_0}{2} + \sum_{k=1}^{\infty} a_k c_k + b_k d_k$$

2.10. Suppose that $0 \le x \le 2\pi$, $0 < d < \pi$, and

$$f(x) = \begin{cases} 1, & 0 \le x \le d \\ 0, & d < x \le 2\pi - d \\ 1, & 2\pi - d < x < 2\pi \end{cases}$$

(a) Compute the Fourier coefficients for f.

(b) Conclude that

$$\sum_{k=1}^{\infty} \frac{\sin kd}{k} = \frac{\pi - d}{2}, \qquad 0 < d < \pi$$

(c) Show that

$$\sum_{k=1}^{\infty} \frac{\sin^2 kd}{k^2 d} = \frac{\pi - d}{2}, \qquad 0 < d < \pi$$

(d) Let $d \to 0$ and prove that

$$\int_0^{\infty} \left(\frac{\sin x}{x} \right)^2 dx = \frac{\pi}{2}$$

(e) Let $d = \pi/2$ in part (c). What do you get?

2.11. To demonstrate the use of Fourier series in differential equations, let f be given by

$$f(x) = \sum_{n=1}^{\infty} b_n \sin nx$$

and assume that

$$\sum_{n=1}^{\infty} |b_n| < \infty$$

Define

$$u(x, t) = \sum_{n=1}^{\infty} b_n \exp\left(-n^2 t\right) \sin nx$$

and prove that

(a) $u(x, t)$ is continuous in the rectangle $0 \le x \le \pi$, $0 \le t \le T$.

(b) the partial derivatives u_x, u_{xx}, u_t exist if $t > 0$.

(c) $u(x, t)$ solves the heat equation $u_{xx} - u_t = 0$, $0 < x < \pi$, $0 < t < T$.

(d) $u(x, 0) = f(x)$, $0 < x < \pi$ and $u(0, t) = u(\pi, t) = 0$, $0 < t < T$.

2.12. (a) Find a sequence of functions f_n, and a function f, so that f_n converges to f pointwise, but not in mean square.

(b) Find a sequence of functions f_n, and a function f, so that f_n converges to f in mean square, but not pointwise.

2.13. Show the Cauchy–Schwarz inequality

$$\sum_{k=1}^{\infty} |a_k b_k| \le \left(\sum_{k=1}^{\infty} a_k^2\right)^{1/2} \left(\sum_{k=1}^{\infty} b_k^2\right)^{1/2}$$

for real a_k, b_k, by considering the polynomial

$$\sum_{k=1}^{\infty} (|a_k| + x|b_k|)^2$$

which can have at most one real root.

2.14. Show Bessel's inequality

$$\frac{a_0^2}{2} + \sum_{k=1}^{\infty} a_k^2 + b_k^2 \le \frac{1}{\pi} \int_0^{2\pi} f^2(t)\, dt$$

by considering

$$\frac{1}{\pi} \int_0^{2\pi} \left(f(t) - \frac{a_0}{2} + \sum_{k=1}^{n} a_k \cos kt + b_k \sin kt\right)^2 dt \ge 0$$

2.15. A sequence of real numbers S_n is said to be a Cauchy sequence if for $m > n$,

$$\lim_{n \to \infty} |S_n - S_m| = 0$$

uniformly in m. This means that for n large enough the difference is uniformly close to 0 regardless of the value of m. Show that every Cauchy sequence converges. Show that if the sequence is a function of x, $S_n = S_n(x)$, and the

Cauchy condition above is also uniform in x, then the sequence converges uniformly in x.

2.16. Suppose that f and f' are continuous periodic functions on $[0, 2\pi]$, and define S_n to be the nth partial sum of the Fourier series for f:

$$S_n = \frac{a_0}{2} + \sum_{k=1}^{n} a_k \cos kt + b_k \sin kt$$

(a) Show $|S_n - S_m| \leq \sum_{k=n}^{\infty} |a_k| + |b_k|$.

(b) Integrate by parts to show that

$$|S_n - S_m| \leq \sum_{k=n}^{\infty} \frac{1}{k}(|c_k| + |d_k|)$$

where

$$c_k = \int_0^{2\pi} f'(t) \sin kt \, dt$$

$$d_k = \int_0^{2\pi} f'(t) \cos kt \, dt$$

(c) Apply the Cauchy–Schwarz inequality to show that

$$|S_n - S_m|^2 \leq \left(\sum_{k=n}^{\infty} |c_k|^2 + |d_k|^2 \right)\left(\sum_{k=n}^{\infty} \frac{2}{k^2} \right)$$

and conclude by Bessel's inequality, for sufficiently large n, that

$$|S_n - S_m| \leq \frac{1}{\sqrt{n}} \int_0^{2\pi} f^2(t) \, dt$$

which implies that the series converges uniformly.

2.17. Let D_n be the Dirichlet kernel

$$D_n(t) = \sum_{|k| \leq n} e^{ikt}$$

Show that

$$D_n(t) = \frac{\sin (n + \frac{1}{2})t}{\sin (\frac{1}{2}t)}$$

and

$$\int_0^{2\pi} D_n(t)\, dt \;=\; 2\pi$$

2.18. Show, with D_n as in Problem 2.17 and f, f' as in Problem 2.16, that the partial sums of the Fourier series for f can be expressed

$$S_n \;=\; \frac{1}{2\pi}\int_0^{2\pi} f(t)D_n(x-t)\, dt$$

which by periodicity and since D_n is even can be written

$$=\; \frac{1}{2\pi}\int_0^{2\pi} f(x-t)D_n(t)\, dt$$

2.19. Show, under the conditions of Problem 2.18, that

$$S_n - f(t) \;=\; \frac{1}{2\pi}\int_0^{2\pi} [f(x-t) - f(x)]D_n(t)\, dt$$

and then, by recognizing the integral as Fourier coefficients of a square-integrable function and applying Bessel's inequality, conclude that S_n converges to $f(t)$ for each t.

CHAPTER

3

Poisson Processes

Informally, a stochastic process is a random quantity that evolves over time. Examples of this phenomenon include the total number of emissions of a radioactive substance measured from a given time, a gambler's net fortune, the position of a particle undergoing some random motion, and the number of bacteria in a culture. Formally, a stochastic process is a family of random variables X_t, where t ranges over a given parameter set T. The index set can correspond to discrete units of time $T = \{1, 2, \ldots\}$ or to a continuous time evolution $T = \mathbb{R}$ or $\mathbb{R}^+$. We shall usually write $X(t)$ for X_t to stress that we are considering the random variables X_t as a function of t. We begin our study of stochastic processes with the Poisson process. In this and later chapters we shall see that many processes are closely related to the Poisson process.

The Poisson process arises naturally in modeling the number of occurrences of independent random events. For example, one may examine, as a function of elapsed time, the number of cars passing a certain point on a highway, the number of particles emitted by a radioactive substance, or the occurrence of accidents, errors, and breakdowns in a variety of situations. In each case the number of occurrences of the event in a time interval $\Delta = (t, t + \Delta t)$ is, on the average, the product of Δt with a fixed number ν which is interpreted as the average frequency of occurrence for the event. Since in small intervals we expect to find at most one occurrence, the probability of exactly one occurrence in a small interval of size Δt is approximately $\nu \, \Delta t$. If the further assumption is made that various occurrences are unrelated, we have what is known as a Poisson process.

3.1   INTRODUCTION TO THE POISSON PROCESS

Let $N(t)$ be the number of occurrences of a random event in the time interval $(0, t]$. If $N(t)$ satisfies the following two assumptions, we shall call $N(t)$ a Poisson process of parameter v.

(i)   There exists $v > 0$ such that for any $t \geq 0$ and $\Delta t > 0$,

(a) $P[N(t + \Delta t) - N(t) = 1] = v \, \Delta t + o(\Delta t)$

(b) $P[N(t + \Delta t) - N(t) = 0] = 1 - v \, \Delta t + o(\Delta t)$ (3.1.1)

(c) $P[N(t + \Delta t) - N(t) = k] = o(\Delta t), \quad k \geq 2$

The notation $o(\Delta t)$ indicates that the quantity it replaces is of higher order than Δt, that is,

$$\frac{o(\Delta t)}{\Delta t} \to 0 \qquad \text{as } \Delta t \to 0 \qquad (3.1.2)$$

(ii)   If N_Δ is the number of occurrences in a time interval $\Delta = (t, t + \Delta t]$, then $N_{\Delta_1}, N_{\Delta_2}, \ldots, N_{\Delta_k}$ are independent whenever $\Delta_1, \Delta_2, \ldots, \Delta_k$ are disjoint. For example, $N(t_1 + \Delta t_1) - N(t_1)$ and $N(t_2 + \Delta t_2) - N(t_2)$ are independent if $(t_1, t_1 + \Delta t_1]$ and $(t_2, t_2 + \Delta t_2]$ do not overlap.

To find the probability distribution of $N(t)$, we introduce the notation

$$P_k(t) = P[N(t) = k] \qquad (3.1.3)$$

so that by assumptions (i) and (ii),

$$P_0(t + \Delta t) = P[N(t + \Delta t) = 0]$$

$$= P[N(t) = 0, N(t + \Delta t) - N(t) = 0]$$

$$= P[N(t) = 0]P[N(t + \Delta t) - N(t) = 0]$$

$$= P_0(t)[1 - v\Delta t + o(\Delta t)]$$

Therefore,

$$\frac{P_0(t + \Delta t) - P_0(t)}{\Delta t} = -vP_0(t) + \frac{o(\Delta t)}{\Delta t} P_0(t)$$

which, as $\Delta t \to 0$, yields

$$\frac{dP_0(t)}{dt} = -vP_0(t)$$

The solution to this differential equation, noting $P_0(0) = 1$, is

$$P_0(t) = e^{-vt}$$

or

$$P[N(t) = 0] = e^{-vt} \qquad (3.1.4)$$

To obtain the general formula

$$P[N(t) = k] = \frac{(vt)^k e^{-vt}}{k!} \qquad (3.1.5)$$

we first compute

$$P_k(t + \Delta t) = P[N(t + \Delta t) = k]$$

$$= \sum_{i=0}^{k} P[N(t) = k - i, N(t + \Delta t) - N(t) = i]$$

which by assumption (ii)

$$= \sum_{i=0}^{k} P[N(t) = k - i]P[N(t + \Delta t) - N(t) = i]$$

$$= \sum_{i=0}^{k} P_{k-i}(t)P[N(t + \Delta t) - N(t) = i]$$

Rewriting the sum, with $\Delta = (t, t + \Delta t]$, gives

$$P_k(t + \Delta t) = P_k(t)P[N_\Delta = 0] + P_{k-1}(t)P[N_\Delta = 1] + \sum_{i=2}^{k} P_{k-i}(t)P[N_\Delta = i]$$

which by assumption (i)

$$= P_k(t)[1 - v\,\Delta t + o(\Delta t)] + P_{k-1}(t)[v\,\Delta t + o(\Delta t)] + \sum_{i=2}^{k} P_{k-i}(t)o(\Delta t)$$

This, upon rearranging and dividing by Δt, becomes

$$\frac{P_k(t + \Delta t) - P_k(t)}{\Delta t} = v(P_{k-1}(t) - P_k(t)) + \frac{o(\Delta t)}{\Delta t}$$

or as $\Delta t \to 0$,

$$\frac{dP_k(t)}{dt} = v(P_{k-1}(t) - P_k(t)) \qquad (3.1.6)$$

For $k = 1$ this is

$$\frac{dP_1(t)}{dt} = v(P_0(t) - P_1(t))$$

which can be solved by using (3.1.4) and noting that $P_1(0) = 0$ to obtain

$$P_1(t) = vte^{-vt}$$

or

$$P[N(t) = 1] = vte^{-vt} \qquad (3.1.7)$$

The general formula (3.1.5) follows by induction.

3.2   AN EQUIVALENT DERIVATION

Once again we calculate $P_k(t)$, but by a method that does not require solving a differential equation. Divide the interval $(0, t]$ into M subintervals

$$\Delta_i = \left(\frac{(i-1)t}{M}, \frac{it}{M} \right], \qquad i = 1, 2, 3, \ldots, M$$

so that

$$P_0(t) = P[N(t) = 0]$$
$$= P[N_{\Delta_1} = 0, N_{\Delta_2} = 0, \ldots, N_{\Delta_M} = 0]$$

which by assumption (ii)

$$= P[N_{\Delta_1} = 0]P[N_{\Delta_2} = 0] \cdots P[N_{\Delta_M} = 0]$$

and by assumption (i)

$$= [1 - v\,\Delta t + o(\Delta t)]^M$$

or, noting $\Delta t = t/M$,

$$= [1 - \nu \, \Delta t + o(\Delta t)]^{t/\Delta t}$$

Therefore,

$$\ln P_0(t) = \frac{t}{\Delta t} \ln [1 - \nu \, \Delta t + o(\Delta t)]$$

and applying the Taylor expansion

$$\ln (1 + x) = x - \frac{x^2}{2} + \frac{x^3}{3} - \frac{x^4}{4} + \cdots \qquad (3.2.1)$$

yields, as $\Delta t \to 0$,

$$P_0(t) = e^{-\nu t}$$

The formula for $P_k(t)$ can be found by similar means, but care must be taken to assure that the $o(\Delta t)$ terms may be neglected. If we assume that this is the case, and take for purposes of calculation

$$(a) \; P[N_{\Delta_i} = 1] = \nu \, \Delta t$$

$$(b) \; P[N_{\Delta_i} = 0] = 1 - \nu \, \Delta t \qquad (3.2.2)$$

$$(c) \; P[N_{\Delta_i} = k] = 0, \quad k \geq 2$$

the result follows from the limiting behavior of the binomial distribution.

3.3   AN ALTERNATIVE DEFINITION OF THE POISSON PROCESS

We have defined the Poisson process $N(t)$ as a process that counts the number of occurrences of a random event in $(0, t]$ and satisfies assumptions (i) and (ii). From this we were able to deduce the distribution of $N(t)$ at any time t. In fact, since zero does not play a role in our assumptions, we conclude that $N_\Delta = N(t + \Delta t) - N(t)$ satisfies

$$(i') \; P[N_\Delta = k] = \frac{(\nu \, \Delta t)^k e^{-\nu \Delta t}}{k!} \qquad \text{for any } \Delta = (t, t + \Delta t] \subset (0, \infty)$$

It is easy to check by considering the Taylor expansion of $e^{-\nu \Delta t}$ that (i') implies (i), so this and the assumption of independence (ii′) $N_{\Delta_1}, N_{\Delta_2}, \ldots, N_{\Delta_n}$ are independent if $\Delta_1, \Delta_2, \ldots, \Delta_n$ are disjoint

are equivalent to assumptions (i) and (ii) in the definition of the Poisson process.

3.4 POISSON PROCESS FROM EXPONENTIAL RANDOM VARIABLES

Recall an exponential random variable X of parameter v has density function

$$f_X(x) = \begin{cases} ve^{-vx}, & x \geq 0 \\ 0, & x < 0 \end{cases}$$

Let X_i, $i = 1, 2, \ldots$ be an independent identically distributed family of random variables with this density. Define the sums S_n by

$$S_1 = X_1$$
$$S_2 = X_1 + X_2$$
$$\vdots \qquad\qquad\qquad (3.4.1)$$
$$S_k = X_1 + X_2 + \cdots + X_k$$
$$\vdots$$

and let $N(t)$ be the number of S_n's that are less than or equal to t. In other words, $N(t) = k$ if and only if

$$X_1 + \cdots + X_k = S_k \leq t < S_{k+1} = X_1 + \cdots + X_{k+1}$$

Here $N(t)$ counts the number of occurrences of a random event, where the time between the kth and $(k + 1)$st occurrences is exponentially distributed, so the process is said to have exponential interarrival times. We shall show that $N(t)$ is a Poisson process by verifying assumptions (i′) and (ii′) of the preceding section.

First we show (i′) by verifying both

$$P[N(t) = k] = \frac{(vt)^k e^{-vt}}{k!} \qquad\qquad (3.4.2)$$

and for $t > r$,

$$P[N(r) = m, N(t) - N(r) = k] = \frac{(vr)^m e^{-vr}}{m!} \frac{[v(t - r)]^k e^{-v(t-r)}}{k!} \qquad (3.4.3)$$

since these two together imply

$$P[N(t) - N(r) = k \mid N(r) = m] = \frac{[v(t-r)]^k e^{-v(t-r)}}{k!}$$

which implies (i'). Define s_n to be the sum $s_n = x_1 + x_2 + \cdots + x_n$. Since the X_i's are independent they have joint density

$$f_X(x_1, \ldots, x_n) = v e^{-vx_1} v e^{-vx_2} \cdots v e^{-vx_n}$$

$$= v^n e^{-vs_n}$$

Using this,

$$P[N(t) = k] = P[S_k \le t < S_{k+1}]$$

$$= \int \cdots \int_{\substack{x_i \ge 0 \\ s_k \le t < s_{k+1}}} v^{k+1} e^{-vs_{k+1}} \, dx_{k+1} \cdots dx_1$$

Noting that the condition $t < s_{k+1}$ is the same as $t - s_k < x_{k+1}$, and integrating over x_{k+1} gives

$$P[N(t) = k] = \int \cdots \int_{\substack{x_i \ge 0 \\ s_k \le t}} v^k e^{-vs_k} \int_{t-s_k}^{\infty} v e^{-vx_{k+1}} \, dx_{k+1} \, dx_k \cdots dx_1$$

$$= v^k e^{-vt} \int \cdots \int_{\substack{x_i \ge 0 \\ s_k \le t}} dx_k \cdots dx_1$$

Since we know from (0.7.1),

$$\int \cdots \int_{\substack{x_i \ge 0 \\ s_k \le t}} dx_k \cdots dx_1 = \frac{t^k}{k!} \tag{3.4.4}$$

it follows that

$$P[N(t) = k] = \frac{(vt)^k e^{-vt}}{k!}$$

which verifies (3.4.2). To show (3.4.3), we write

$$P[N(r) = m, N(t) - N(r) = k]$$
$$= P[N(r) = m, N(t) = m + k]$$
$$= P[S_m \leq r < S_{m+1}, S_{k+m} \leq t < S_{k+m+1}]$$
$$= \int \cdots \int_{\substack{x_i \geq 0 \\ S_m \leq r < S_{m+1} \\ S_{k+m} \leq t < S_{k+m+1}}} e^{-vs_{k+m+1}} v^{k+m+1}\, dx_{k+m+1} \cdots dx_1$$

which when integrated over x_{k+m+1}

$$= \int \cdots \int_{\substack{x_i \geq 0 \\ S_m \leq r < S_{m+1} \\ S_{k+m} \leq t}} v^{k+m} e^{-vs_{k+m}} \int_{t-S_{k+m}}^{\infty} v e^{-vx_{k+m+1}}\, dx_{k+m+1} \cdots dx_1$$

or

$$= \int \cdots \int_{\substack{x_i \geq 0 \\ S_m \leq r < S_{m+1} \\ S_{k+m} \leq t}} v^{k+m} e^{-vt}\, dx_{k+m} \cdots dx_1$$

This can be written

$$P[N(r) = m, N(t) - N(r) = k]$$
$$= \int \cdots \int_{\substack{x_i \geq 0 \\ S_m \leq r}} v^{k+m} e^{-vt} \left[\int_{r-S_m}^{t-S_m} \int_0^{t-S_{m+1}} \cdots \int_0^{t-S_{k+m-1}} dx_{k+m} \cdots dx_{k+1} \right] dx_k \cdots dx_1$$

which is

$$= \int \cdots \int_{\substack{x_i \geq 0 \\ S_m \leq r}} v^{k+m} e^{-vt} \frac{(t-r)^k}{k!}\, dx_m \cdots dx_1$$

and by (3.4.4)

$$= \frac{(vr)^m e^{-vr}}{m!} \frac{[v(t-r)]^k e^{-v(t-r)}}{k!}$$

which is (3.4.3). Thus (i′) is true.

To show (ii′), we must verify that

$$P[N_{\Delta_1} = k_1, \ldots, N_{\Delta_n} = k_n] = P[N_{\Delta_1} = k_1]P[N_{\Delta_2} = k_2] \cdots P[N_{\Delta_n} = k_n] \tag{3.4.5}$$

holds for disjoint intervals Δ_i. We show this first for a particular pair of adjacent intervals and then for arbitrary intervals. Suppose that Δ_1, Δ_2 are the adjacent intervals $\Delta_1 = (0, r]$ and $\Delta_2 = (r, t]$; then by (3.4.3)

$$
\begin{aligned}
P[N_{\Delta_1} &= m, N_{\Delta_2} = k] \\
&= P[N(r) = m, N(t) - N(r) = k] \\
&= \frac{(vr)^m e^{-vr}}{m!} \frac{[v(t - r)]^k e^{-v(t-r)}}{k!} \\
&= P[N_{\Delta_1} = m]P[N_{\Delta_2} = k]
\end{aligned}
$$

so we have shown the independence for a pair of adjacent intervals, where one begins at 0. The proof of the formula (3.4.5) for an arbitrary number of adjacent intervals Δ_i whose union is the interval $(0, t]$ is left to the reader.

For an arbitrary pair of disjoint nonadjacent intervals $\Delta_1 = (s, s + \Delta s]$ and $\Delta_2 = (t, t + \Delta t), 0 < s < s + \Delta s < t < t + \Delta t$, add to the collection the intervals $\Delta_0 = (0, s]$ and $\Delta_3 = (s + \Delta s, t]$, so that we may conclude that

$$
\begin{aligned}
P[N_{\Delta_0} &= m, N_{\Delta_1} = h, N_{\Delta_2} = k, N_{\Delta_3} = n] \\
&= P[N_{\Delta_0} = m]P[N_{\Delta_1} = h]P[N_{\Delta_2} = k]P[N_{\Delta_3} = n]
\end{aligned} \tag{3.4.7}
$$

Summing the above over all possible values of m and n yields

$$P[N_{\Delta_1} = h, N_{\Delta_2} = k] = P[N_{\Delta_1} = h]P[N_{\Delta_2} = k]$$

as desired. The corresponding result for a larger number of nonadjacent intervals follows the same line of reasoning. From this we conclude (ii′).

3.5   POISSON PROCESS FROM UNIFORM RANDOM VARIABLES

Let s be a fixed positive number, $T_i, i = 1, 2, \ldots$ be independent identically distributed random variables uniform on $(0, s]$ and N a Poisson random variable satisfying

$$P[N = k] = \frac{(vs)^k e^{-vs}}{k!}$$

independent of the T_i's. Let $1_{(a,b]}$ be the indicator function of an interval $(a, b]$,

$$1_{(a,b]}(t) = \begin{cases} 1, & t \in (a, b] \\ 0, & \text{else} \end{cases} \qquad (3.5.1)$$

and define $N(t)$ by

$$N(t) = \sum_{j=1}^{N} 1_{(0,t]}(T_j), \qquad t < s \qquad (3.5.2)$$

so $N(t)$ counts the number of occurrences in $(0, t]$ created by taking one sample from each of N independent uniform distributions on the larger interval $(0, s]$. We now prove that this is a description of the Poisson process during time $(0, s]$. What we shall show is that (i') and (ii') are satisfied.

To show (i') we use the Fourier inversion formula. Letting $\Delta = (t, t + \Delta t]$, we first find the characteristic function of N_Δ,

$$E[e^{i\xi N_\Delta}] = E\left[\exp\left(i\xi \sum_{j=1}^{N} 1_\Delta(T_j) \right) \right] \qquad (3.5.3)$$

$$= \sum_{k=0}^{\infty} E\left[\exp\left(i\xi \sum_{j=1}^{k} 1_\Delta(T_j) \right) \right] P[N = k]$$

which since the T_i's are independent identically distributed,

$$= \sum_{k=0}^{\infty} \left(E[\exp(i\xi 1_\Delta(T_1))] \right)^k \frac{(vs)^k e^{-vs}}{k!}$$

$$= e^{-vs} e^{vs E[\exp(i\xi 1_\Delta(T_1))]}$$

The value of the expectation

$$E[\exp(i\xi 1_\Delta(T_1))] = \int_0^s e^{i\xi 1_\Delta(t)} \frac{1}{s} dt$$

from the definition of 1_Δ is

$$= \int_\Delta e^{i\xi} \frac{1}{s} dt + \int_{(0,s]\backslash\Delta} \frac{1}{s} dt$$

$$= \frac{\Delta t e^{i\xi} + s - \Delta t}{s}$$

Substituting this in (3.5.3) we obtain

$$E[e^{i\xi N_\Delta}] = e^{\nu\Delta t(e^{i\xi}-1)} \tag{3.5.4}$$

Applying the Fourier inversion formula to this reveals the density function of N_Δ to be

$$\frac{1}{2\pi}\int_{-\infty}^{\infty} e^{\nu\Delta t(e^{i\xi}-1)}e^{-i\xi x}\,d\xi = \frac{1}{2\pi}\int_{-\infty}^{\infty} e^{-\nu\Delta t}\left[\sum_{k=0}^{\infty}(\nu\Delta t)^k e^{i\xi k}\frac{1}{k!}\right]e^{-i\xi x}\,d\xi$$

$$= \sum_{k=0}^{\infty} e^{-\nu\Delta t}(\nu\,\Delta t)^k\frac{1}{k!}\frac{1}{2\pi}\int_{-\infty}^{\infty} e^{i\xi(k-x)}\,d\xi$$

$$= \sum_{k=0}^{\infty} e^{-\nu\Delta t}(\nu\,\Delta t)^k\frac{1}{k!}\delta(x-k)$$

This says that N_Δ satisfies

$$P[N_\Delta = k] = \frac{(\nu\,\Delta t)^k e^{-\nu\Delta t}}{k!} \tag{3.5.5}$$

as required. We now show the independence (ii'), for two arbitrary disjoint intervals, after which we suggest the proof for the general case.

Suppose that $\Delta_1 = (t_1, t_1 + \Delta t_1]$ and $\Delta_2 = (t_2, t_2 + \Delta t_2]$ are two disjoint intervals contained in $(0, s]$; then for any real value of ξ and η,

$$E[\exp(i(\xi N_{\Delta_1} + \eta N_{\Delta_2}))] = E[\exp(i\xi N_{\Delta_1})]E[\exp(i\eta N_{\Delta_2})],$$

which says the two random variables N_{Δ_1} and N_{Δ_2} are independent. The proof of this is as follows:

$$E[\exp(i(\xi N_{\Delta_1} + \eta N_{\Delta_2}))]$$

$$= E\left[\exp\left(i\xi\sum_{j=1}^{N}1_{\Delta_1}(T_j) + i\eta\sum_{j=1}^{N}1_{\Delta_2}(T_j)\right)\right]$$

$$= \sum_{k=0}^{\infty}E\left[\exp\left(i\xi\sum_{j=1}^{k}1_{\Delta_1}(T_j) + i\eta\sum_{j=1}^{k}1_{\Delta_2}(T_j)\right)\right]P[N=k] \tag{3.5.6}$$

$$= \sum_{k=0}^{\infty}(E[\exp(i\xi 1_{\Delta_1}(T_1) + i\eta 1_{\Delta_2}(T_1))])^k\frac{e^{-\nu s}(\nu s)^k}{k!}$$

$$= e^{-\nu s}\sum_{k=0}^{\infty}\frac{(\nu s E[\exp(i\xi 1_{\Delta_1}(T_1) + i\eta 1_{\Delta_2}(T_1))])^k}{k!}$$

$$= e^{-\nu s}e^{\nu s E[\exp(i\xi 1_{\Delta_1}(T_1) + i\eta 1_{\Delta_2}(T_1))]}$$

We evaluate the expectation

$$E[\exp{(i\xi 1_{\Delta_1}(T_1) + i\eta 1_{\Delta_2}(T_1))}] = \int_0^s e^{i\xi 1_{\Delta_1}(t) + i\eta 1_{\Delta_2}(t)} \frac{1}{s} dt$$

using the definition of f_Δ,

$$= \int_{\Delta_1} e^{i\xi} \frac{1}{s} dt + \int_{\Delta_2} e^{i\eta} \frac{1}{s} dt + \int_{(0,s]\setminus(\Delta_1\cup\Delta_2)} \frac{1}{s} dt$$

$$= \frac{1}{s} (\Delta t_1 e^{i\xi} + \Delta t_2 e^{i\eta} + s - \Delta t_1 - \Delta t_2).$$

Substituting this in (3.5.6) we get

$$E[\exp{(i(\xi N_{\Delta_1} + \eta N_{\Delta_2}))}]$$
$$= e^{-vs} e^{vsE[\exp(i\xi 1_{\Delta_1}(T_1) + i\eta 1_{\Delta_2}(T_1))]}$$
$$= e^{-vs} e^{v(\Delta t_1 e^{i\xi} + \Delta t_2 e^{i\eta} + s - \Delta t_1 - \Delta t_2)} \qquad (3.5.7)$$
$$= e^{v\Delta t_1(e^{i\xi}-1)} e^{v\Delta t_2(e^{i\eta}-1)}$$

which by (3.5.4) is

$$= E[\exp{(i\xi N_{\Delta_1})}] E[\exp{(i\eta N_{\Delta_2})}]$$

The demonstration of the independence for more than two intervals is similar.

3.6 POISSON PROCESS AND THE ORDER STATISTICS

Suppose that $U_1, \ldots, U_k$ are independent identically distributed random variables uniform on $[0, s]$, and let $U_{(1)} \leq U_{(2)} \leq \cdots \leq U_{(k)}$ be the order statistics of the U_i's, so that $U_{(1)} = \min{(U_i)}$, $U_{(i)} = i$th smallest of the U_i's, and $U_{(k)} = \max{(U_i)}$. The joint distribution of the order statistics can be found by combining the results of the preceding two sections. By Section 3.5, conditioned on $N(s) = k$, the times of occurrence of the events in a Poisson process come from a sample of k independent random variables uniform on $[0, s]$, which means that

$$P[N(t) = m | N(s) = k] = P[U_{(m)} \leq t < U_{(m+1)}]$$

More generally, using the notation of Section 3.4,

$$P[S_1 \leq t_1, \ldots, S_k \leq t_k | N(s) = k] = P[U_{(1)} \leq t_1, \ldots, U_{(k)} \leq t_k] \quad (3.6.1)$$

where S_i, $i = 1, 2, \ldots, k$, are the times of occurrence of the events, and $S_1 \leq S_2 \leq \cdots \leq S_k$. We can evaluate the left side of this explicitly. First noting that

$$P[S_1 \leq t_1, \ldots, S_k \leq t_k | N(s) = k] = \frac{P[S_1 \leq t_1, \ldots, S_k \leq t_k, N(s) = k]}{P[N(s) = k]}$$

and then using the fact that the interarrival times are independent and exponentially distributed, and recalling (3.4.1), give

$$P[S_1 \leq t_1, \ldots, S_k \leq t_k, N(s) = k]$$

$$= P[X_1 \leq t_1, \ldots, X_1 + \cdots + X_k \leq t_k, N(s) = k]$$

$$= \int_0^{t_1} \int_0^{t_2 - x_1} \cdots \int_0^{t_k - s_{k-1}} \int_{s - s_k}^{\infty} v^{k+1} e^{-v s_{k+1}} \, dx_{k+1} \cdots dx_1$$

$$= \int_0^{t_1} \int_0^{t_2 - x_1} \cdots \int_0^{t_k - s_{k-1}} v^k e^{-vs} \, dx_k \cdots dx_1$$

Therefore, using (3.1.5), we obtain

$$P[U_{(1)} \leq t_1, \ldots, U_{(k)} \leq t_k]$$

$$= P[S_1 \leq t_1, \ldots, S_k \leq t_k | N(s) = k] \qquad (3.6.2)$$

$$= \frac{k!}{(vs)^k e^{-vs}} \int_0^{t_1} \int_0^{t_2 - x_1} \cdots \int_0^{t_k - s_{k-1}} v^k e^{-vs} \, dx_k \cdots dx_1$$

$$= \frac{k!}{s^k} \int_0^{t_1} \int_0^{t_2 - x_1} \cdots \int_0^{t_k - s_{k-1}} dx_k \cdots dx_1$$

Differentiating this with respect to $t_1, \ldots, t_k$ shows that the joint density for the order statistics is

$$f(t_1, \ldots, t_k) = \begin{cases} \dfrac{k!}{s^k}, & t_1 \leq t_2 \leq \cdots \leq t_k \leq s \\ 0, & \text{else} \end{cases} \qquad (3.6.3)$$

This is interpreted as saying that the order statistics are uniformly distributed in the region $t_1 \leq t_2 \leq \cdots \leq t_k \leq s$.

3.7   POISSON PROCESS AND THE BINOMIAL DISTRIBUTION

Let $N(t)$ be a Poisson process and s a fixed time. For $0 < t < s$ we may calculate directly

$$P[N(t) = 1|N(s) = k] = \frac{[N(t) = 1, N(s) = k]}{P[N(s) = k]}$$

$$= \frac{P[N(t) = 1, N(s) - N(t) = k - 1]}{P[N(s) = k]}$$

$$= \frac{P[N(t) = 1]P[N(s) - N(t) = k - 1]}{P[N(s) = k]}$$

$$= \frac{[(vt)e^{-vt}][v(s - t)^{k-1}e^{-v(s-t)}/(k - 1)!]}{(vs)^k e^{-vs}/k!}$$

$$= k\left(\frac{t}{s}\right)\left(1 - \frac{t}{s}\right)^{k-1}$$

and similarly,

$$P[N(t) = m|N(s) = k] = \frac{P[N(t) = m]P[N(s) - N(t) = k - m]}{P[N(s) = k]}$$

$$= \binom{k}{m}\left(\frac{t}{s}\right)^m\left(1 - \frac{t}{s}\right)^{k-m}$$

The binomial character of these probabilities is no accident. We know that the Poisson process is related to the order statistics by the formula

$$P[U_{(1)} \leq t, U_{(2)} > t] = P[N(t) = 1|N(s) = k]$$

and in general

$$P[U_{(m)} \leq t, U_{(m+1)} > t] = P[N(t) = m|N(s) = k] \tag{3.6.1}$$

which has a binomial distribution. Of course, the situation is even more general than this. Suppose that $\Delta = (t, t + \Delta t]$ is contained in $(0, s]$. By Section 3.4, if we know the number of occurrences in $(0, s]$, the times of these occurrences T_i are independent and uniformly distributed. Therefore, $P[N_\Delta = m|N(s) = k]$ is the probability of m successes in k binomial trials, where the probability of success is

$$P[T_i \in \Delta] = \frac{\Delta t}{s}$$

implying that

$$P[N_\Delta = m|N(s) = k] = \binom{k}{m}\left(\frac{\Delta t}{s}\right)^m\left(1 - \frac{\Delta t}{s}\right)^{k-m} \tag{3.6.2}$$

3.8   GENERALIZATIONS OF THE POISSON PROCESS

We have seen various formulations of the Poisson process, which is based on the assumption that $N(t)$ counts the number of occurrences of a random event and satisfies

(i)   There exists a parameter $v > 0$ such that
     (a) $P[N(t + \Delta t) - N(t) = 1] = v \, \Delta t + o(\Delta t)$
     (b) $P[N(t + \Delta t) - N(t) = 0] = 1 - v \, \Delta t + o(\Delta t)$
     (c) $P[N(t + \Delta t) - N(t) = k] = o(\Delta t)$

(ii)   The increments

$$N_{\Delta_i} = N(t_i + \Delta t_i) - N(t_i)$$

are independent if the intervals $\Delta_i = (t_i, t_i + \Delta_i]$ are disjoint.

The parameter v is interpreted as the average rate of occurrences of the random event.

These assumptions are appropriate for radioactive emissions over the short run, but over the long run the rate of radioactive emissions decreases, due to the decreased amount of material subject to radioactive decay. To take this phenomenon into account, one can let the rate v be a function of the number of occurrences that have taken place, and in place of (i) assume:

(i) There exist parameters $v_i \geq 0$ such that

(a) $P[N(t + \Delta t) - N(t) = 1 | N(t) = i] = v_i \, \Delta t + o(\Delta t)$

(b) $P[N(t + \Delta t) - N(t) = 0 | N(t) = i] = 1 - v_i \, \Delta t + o(\Delta t)$ (3.8.1)

(c) $P[N(t + \Delta t) - N(t) = k | N(t) = i] = o(\Delta t)$

Since the rate depends on how many occurrences have taken place, the number of occurrences in disjoint intervals need not be independent. For example, the value of $N(t) - N(0)$ affects the rate of occurrences over $(t, t + s]$, so it affects $N(t + s) - N(t)$. However, since the dependence is completely determined by value of $N(t)$, the independence assumption can be replaced by a conditional independence assumption:

(ii)   Given $N(t)$, $N_\Delta = N(t + \Delta t) - N(t)$ is independent of the family $\{N_{\Delta_i}\}$ where $N_{\Delta_i} = N(t_i + \Delta t_i) - N(t_i)$ and the intervals $\Delta_i = (t_i, t_i + \Delta_i]$ are contained in $(0, t]$.

In this case we have defined what is known as a pure birth process. Condition (ii) is a form of the **Markov property,** which states that the future evolution of the process depends on the "past" only through its "present" value. We shall see the Markov property again in Chapter 7. As in the Poisson case, a pure birth process can be described in terms of a sequence X_i of independent exponential random variables, but this time of varying parameters v_i.

Let us find the distribution of $N(t)$ under these new assumptions. Proceeding as we did for the Poisson process at the beginning of the chapter, we obtain

$$P[N(t) = 0] = e^{-\nu_0 t}$$

Further, setting

$$P_k(t) = P[N(t) = k]$$

gives

$$P_k(t + \Delta t) = P_k(t)P[N_\Delta = 0 | N(t) = k]$$
$$+ P_{k-1}(t)P[N_\Delta = 1 | N(t) = k - 1]$$
$$+ \sum_{i=2}^{k} P_{k-i}(t)P[N_\Delta = i | N(t) = k - i]$$

which becomes

$$P_k(t + \Delta t) = P_k(t)[1 - \nu_k \Delta t + o(\Delta t)] + P_{k-1}(t)[\nu_{k-1} \Delta t + o(\Delta t)]$$
$$+ \sum_{i=2}^{k} P_{k-i}(t)o(\Delta t)$$

so that

$$\frac{dP_k(t)}{dt} = \nu_{k-1}P_{k-1}(t) - \nu_k P_k(t) \tag{3.8.2}$$

This system of differential equations can be solved for any choice of $\nu_i \geq 0$, with initial condition $P_k(t) = 0$, $k > 0$ yielding the recursion relation

$$P_k(t) = e^{-\nu_k t} \int_0^t e^{\nu_k s} \nu_{k-1} P_{k-1}(s) \, ds \tag{3.8.3}$$

which allows the calculation of $P_k(t)$ for any k.

Notice that the rates ν_k only depend on the number of occurrences and not on the elapsed time. Thus for a fixed time s, the behavior of the process at time $s + t$ depends only on $N(s)$. This means that the function

$$p_{ij}(t) = P[N(t + s) = j | N(s) = i]$$

does not depend on s. The quantity $p_{ij}(t)$ is called the transition probability function of the process, since it tells the probability that in time t we see the number of occurrences increase from i to j, or that the "state" of the process changes from $N(s) = i$ to $N(t + s) = j$. Since $p_{ij}(t)$ is independent of s, the process is said to have **stationary** transition probabilities.

We can get a differential equation for the transition probabilities by writing

$$p_{ij}(t + \Delta t) = p_{ij}(t)P[N_\Delta = 0|N(t) = j] + p_{ij-1}(t)P[N_\Delta = 1|N(t) = j - 1]$$

$$+ \sum_{k=2}^{\infty} p_{ij-k}(t)P[N_\Delta = k|N(t) = j - k]$$

and obtain

$$\frac{d(p_{ij}(t))}{dt} = v_{j-1}p_{ij-1}(t) - v_j p_{ij}(t)$$

which can be solved in the same way as (3.8.2). This is called the **forward system of equations,** and it involves the transition probabilities for a fixed initial state i. Intuitively this equation says the probability of being at state j grows at a rate at which is equal to the rate at which the process enters the state j minus the rate at which it leaves state j.

There is another system of differential equations associated with this process, obtained by comparing $p_{ij}(\Delta t)$ and $p_{ij}(t + \Delta t)$:

$$p_{ij}(t + \Delta t) = p_{ij}(\Delta t)P[N(t + \Delta t) - N(\Delta t) = 0|N(\Delta t) = j]$$

$$+ p_{ij-1}(\Delta t)P[N(t + \Delta t) - N(\Delta t) = 1|N(\Delta t) = j - 1]$$

$$+ \sum_{k=2}^{\infty} p_{ij-k}(\Delta t)P[N(t + \Delta t) - N(\Delta t) = k|N(\Delta t) = j - k]$$

which is

$$p_{ij}(t + \Delta t) = p_{ij}(\Delta t)p_{jj}(t) + p_{ij-1}(\Delta t)p_{ij-1}(t) + \sum_{k=2}^{\infty} p_{ij-k}(\Delta t)p_{jj-k}'(t)$$

This implies that the transition probabilities also satisfy the differential equations

$$\frac{d(p_{ij}(t))}{dt} = v_i p_{i+1j}(t) - v_i p_{ij}(t)$$

which are called the **backward system of equations.**

Explosions

For the pure birth process, with the appropriate choice of the v_i's it is possible that the number of occurrences approaches infinity in a finite time, that is,

$$P[N(t) < \infty] < 1$$

and the process is said to explode. To find when this happens, let X_i be independent and exponentially distributed with density

$$f_i(x) = e^{-\nu_i x}, \; x \geq 0$$

Then S_n given by

$$S_n = \sum_{k=1}^{n} X_k$$

is the time of occurrence of the nth event. There will be explosion exactly when

$$P[\lim_{n \to \infty} S_n < \infty] \neq 0$$

or when

$$\lim_{n \to \infty} E[e^{-S_n}] \neq 0$$

However,

$$E[e^{-S_n}] = E\left[e^{-\sum_{k=1}^{n} X_k} \right] = \prod_{k=1}^{n} E[e^{-X_k}]$$

and since X_k is exponential of parameter ν_k,

$$E[e^{-X_k}] = 1 - \frac{1}{\nu_k + 1}$$

and

$$E[e^{-S_n}] = \prod_{k=1}^{n} \left(1 - \frac{1}{\nu_k + 1} \right)$$

which as $n \to \infty$ is zero exactly when

$$\sum_{k=1}^{\infty} \frac{1}{\nu_k} = \infty$$

We conclude that explosions occur whenever the sum above is finite. This seems reasonable, since it is the sum of the means of the individual exponential random variables.

One reason these processes are referred to as birth processes is that they can be used to model population growth. For example, with $\nu_k = k\lambda$, $N(t)$ models a population where each individual has a constant average rate λ of giving birth, and there are no deaths. It is clear that this is an unsatisfactory model in many situations, since "deaths" do play a role. We can modify our assumptions to take this into account by replacing (i) by

(i) There exist parameters $v_i \geq 0$, $\mu_i \geq 0$, $\mu_0 = 0$ such that

(a) $P[N(t + \Delta t) - N(t) = 1 | N(t) = i] = v_i \Delta t + o(\Delta t)$

(b) $P[N(t + \Delta t) - N(t) = -1 | N(t) = i] = \mu_i \Delta t + o(\Delta t)$ (3.8.4)

(c) $P[N(t + \Delta t) - N(t) = 0 | N(t) = i] = 1 - v_i \Delta t - \mu_i \Delta t + o(\Delta t)$

(d) $P[N(t + \Delta t) - N(t) = k | N(t) = i] = o(\Delta t)$

We also assume the births and deaths are independent. A process satisfying this and the independence assumption (ii) is called a birth and death process. The reason that $\mu_0 = 0$ is to prevent the process from taking on negative values.

A matrix formulation of the forward and backward equations

In both the birth process and more general birth and death process, we can benefit from considering the matrix of transition probabilities $\mathbf{P}_t$:

$$\mathbf{P}_t = (p_{ij}(t))$$

As a consequence of the independence (ii) these matrices have the property that they form a **semigroup,** by which we mean

$$\mathbf{P}_{t+s} = \mathbf{P}_t \mathbf{P}_s \qquad (3.8.5)$$

This can easily be seen by the following calculations:

$$p_{ij}(t + s) = P[X(t + s) = j | X(0) = i]$$
$$= \sum_k P[X(t + s) = j, X(t) = k | X(0) = i]$$
$$= \sum_k P[X(t + s) = j | X(t) = k] P[X(t) = k | X(0) = i]$$
$$= \sum_k P[X(s) = j | X(0) = k] P[X(t) = k | X(0) = i]$$
$$= \sum_k p_{kj}(s) p_{ik}(t)$$

Equation (3.8.5) is one form of what is called the **Chapman–Kolmogorov equation.**
    The differentiations of the transition functions have a nice form when written in this matrix formulation:

$$\frac{d\mathbf{P}_t}{dt} = \lim_{\Delta t \to 0}\left(\frac{1}{\Delta t}\,(\mathbf{P}_{t+\Delta t} - \mathbf{P}_t)\right) = \mathbf{P}_t\!\left(\lim_{\Delta t \to 0}\left(\frac{1}{\Delta t}\,(\mathbf{P}_{\Delta t} - \mathbf{I})\right)\right)$$

and thus the differential equations are determined by

$$\mathbf{G} = \lim_{\Delta t \to 0}\left(\frac{1}{\Delta t}\,(\mathbf{P}_{\Delta t} - \mathbf{I})\right)$$

which is called the **generator** of the process. The forward equations now take the form

$$\frac{d\mathbf{P}_t}{dt} = \mathbf{P}_t\mathbf{G}$$

We can also get the backward equations

$$\frac{d\mathbf{P}_t}{dt} = \lim\left(\frac{1}{\Delta t}\,(\mathbf{P}_{\Delta t + t} - \mathbf{P}_t)\right) = \left[\lim\left(\frac{1}{\Delta t}\,(\mathbf{P}_{\Delta t} - \mathbf{I})\right)\right]\mathbf{P}_t = \mathbf{G}\mathbf{P}_t$$

Frequently, subject to the initial condition $\mathbf{P}_0 = \mathbf{I}$ the forward and backward equations have a unique solution which is suggested by the exponential nature of these equations,

$$\mathbf{P}_t = \exp{(t\mathbf{G})}$$

where the exponential function is defined as the infinite sum

$$\exp{(\mathbf{A})} = \sum_{k=0}^{\infty} \frac{1}{k!}\,\mathbf{A}^k$$

There is a little more here than meets the eye, since we are dealing with infinite matrices. The limiting process of taking the derivative is more delicate than we would be led to believe just by looking at the equations. We shall not go into the details here, and conclude with the comment that this approach can be pursued with benefit even for more general stochastic processes.

Renewal processes

We have seen that the Poisson process counts the number of occurrences of a random event, where the interarrival times are independent and identically exponentially distributed. Pure birth processes are generalizations of this process, where the assumption of identical interarrival times is relaxed. The birth and death process, in addition, allows what amounts to "negative" occurrences. One could think of birth and death as modeling the total charge accumulating with the arrival of charged particles, of charges $+1$ and -1.

The assumption of exponential interarrival times is necessary if the resulting process is to be Markovian. Recall that a process is Markovian if for any s, the evolution of the process $X(t)$ for time $t \geq s$, and the evolution of the process over $t \leq s$ are independent, conditioned on the value of $X(s)$. Roughly speaking, what happens depends on the past only through the present. We give a formal definition in Chapter 7.

There are instances where the interarrival times are not exponentially distributed, but still may be regarded as independent. Perhaps the simplest of these is when $N(t)$ models the number of times it is necessary to replace a light bulb in time t. Here a lamp is left on until the bulb burns out, at which time the bulb is replaced and the lamp is left on once more. We can assume that the lifetimes of the bulbs are independent identically distributed. The probability distribution of the lifetime of a light bulb can be determined by experiment and is not exponential.

With this in mind, we define $N(t)$ to be renewal process if

$$N(t) = \max (n : S_n \leq t)$$

$$S_n = X_1 + \cdots + X_n$$

where X_i is a sequence of independent identically distributed nonnegative random variables. By convention $S_0 = 0$.

$N(t)$ is interpreted as the number of occurrences of a random event until time t, where the interarrival times are independent with a common distribution. The term "renewal" refers to the fact that once an occurrence is observed, the process "starts afresh" as far as the next occurrence is concerned.

We first find the distribution of $N(t)$. Let F_k be the distribution function of S_k,

$$F_k(t) = P[S_k \leq t]$$

Then

$$P[N(t) = k] = P[S_k \leq t, S_{k+1} > t]$$
$$= F_k(t) - F_{k+1}(t)$$

Thus the mean of $N(t)$ is

$$E[N(t)] = \sum_{k=1}^{\infty} k(F_k(t) - F_{k+1}(t))$$

$$= \sum_{k=1}^{\infty} F_k(t)$$

For convenience, let us assume that the X_i have a density function $f(x)$ and set

$$F(t) = \int_{-\infty}^{t} f(x)\ dx$$

where $f(x) = 0$ for $x < 0$, since X is positive. Recalling that convolution

$$h * g(t) = \int_{-\infty}^{\infty} h(t - x)g(x)\ dx$$

gives the density for the sum of two independent random variables with densities g and h, we have

$$F_k(t) = \int_{-\infty}^{t} f^{*k}(x)\ dx$$

where f^{*k} denotes the k-fold convolution of f. Thus

$$E[N(t)] = \sum_{k=1}^{\infty} \int_{-\infty}^{t} f^{*k}(x)\ dx$$

Now, defining the renewal function

$$m(t) = E[N(t)]$$

by interchanging the integral and sum, we have

$$m(t) = \int_{-\infty}^{t} \sum_{k=1}^{\infty} f^{*k}(x)\ dx$$

By convolution with f and addition of F on each side,

$$m * f(t) + F(t) = \int_{-\infty}^{t} \sum_{k=2}^{\infty} f^{*k}(x)\ dx + \int_{-\infty}^{t} f(x)\ dx = m(t)$$

we see that $m(t)$ satisfies

$$m(t) = F(t) + m * f(t)$$

Equations of the form

$$m(t) = y(t) + m * f(t)$$

are called renewal equations, and occur frequently in the study of properties of renewal processes. Their strength lies in the fact that when $y(t) = 0$ for $x < 0$, and is bounded, the solution is

$$m(t) = \int_{-\infty}^{\infty} y(t - x)U(x) \, dx$$

or

$$m(t) = y * U(t)$$

where U is the **potential** of S_n, defined as the sum of its density functions for each $n \geq 0$

$$U(t) = \sum_{k=1}^{\infty} f^{*k}(t) + \delta(t)$$

For a renewal process the mean satisfies

$$m(t) = F(t) + m * f(t)$$

so

$$m(t) = \int_{-\infty}^{\infty} F(t - x)U(x) \, dx$$

Suppose that we have a Poisson process, where the interarrival times are exponential with density

$$f_X(x) = \lambda e^{-\lambda x}, \qquad x \geq 0$$

Then U is a sum of gamma densities

$$U(x) = \sum_{k=1}^{\infty} \frac{\lambda e^{-\lambda x}(\lambda x)^{k-1}}{(k - 1)!} + \delta(x)$$

$$= \lambda + \delta(x)$$

and

$$m(t) = \int_{0}^{t} (1 - e^{-\lambda(t-x)})\lambda \, dx + 1 - e^{-\lambda t}$$

$$= \lambda t$$

as we found before.

The theory of renewal processes is primarily concerned with the limiting behavior of the renewal function

$$m(t) = E[N(t)]$$

We cannot go into this in depth, and will be content with showing only the basic result

$$\lim_{t \to \infty} \frac{N(t)}{t} = \frac{1}{E[X_1]}$$

almost surely. To show this, first note by the strong law of large numbers

$$\lim_{n \to \infty} \frac{S_n}{n} = E[X_1]$$

almost surely. We have

$$S_{N(t)} \le t < S_{N(t)+1}$$

which gives

$$\frac{S_{N(t)}}{N(t)} \le \frac{t}{N(t)} < \frac{S_{N(t)+1}}{N(t)+1} \frac{N(t)+1}{N(t)}$$

Now as $t \to \infty$, $N(t) \to \infty$ almost surely, so in the limit

$$E[X_1] \le \lim_{t \to \infty} \frac{t}{N(t)} < E[X_1]$$

almost surely, which completes the result.

PROBLEMS

3.1. Use equation (3.1.6) to show by induction that the general formula (3.1.5) is true.

3.2. Show that if the increments $N(t_i + \Delta t) - N(t_i)$ are independent, then the assumptions of Section 3.2 and equations (3.2.2) imply that

$$P_k(t) = \frac{(vt)^k e^{-vt}}{k!}$$

3.3. Prove that if $\{\Delta_1, \ldots, \Delta_n\}$ is a collection of disjoint intervals whose union is $(0, t]$, then

$$P[N_{\Delta_1} = m_1, \ldots, N_{\Delta_n} = m_n] = P[N_{\Delta_1} = m_1] \cdots P[N_{\Delta_n} = m_n]$$

Hint: We have shown this in the case $n = 2$; do it by induction.

3.4. Show under the assumptions of Section 3.5 that the independence (ii′) is satisfied.

3.5. Calculate the mean function

$$\mu(t) = E[N(t)]$$

and the covariance function

$$r(t, s) = E[(N(t) - \mu(t))(N(s) - \mu(s))]$$

for the Poisson process.

3.6. Calculate the mean and covariance function for

$$Y(t) = N(t) - tN(1)$$

3.7. Suppose that $N_1(t)$, $N_2(t)$ are Poisson processes with densities ν_1, ν_2, respectively. Show that

$$N(t) = N_1(t) + N_2(t)$$

is a Poisson process with density $\nu_1 + \nu_2$. Show that a similar result holds if you have a sum of a finite number of independent Poisson processes. Give a counterexample to show that the assumption of independence is necessary. Is the assumption of the independence of the increments violated?

3.8. Using the characteristic function for a Poisson random variable

$$C_N(\xi) = \exp [\nu(e^{i\xi} - 1)]$$

and the independence of the increments N_Δ, find the joint characteristic function for the Poisson process:

$$C(\xi_1, \ldots, \xi_n) = E[\exp (i(\xi_1 N(t_1) + \cdots + \xi_n N(t_n)))]$$

3.9. If N is a Poisson random variable with density νt, use the description given in (3.5.2) of this random variable to verify that as $\nu \to \infty$,

$$\frac{N - \nu t}{\sqrt{\nu t}}$$

becomes normal with mean zero and variance 1.

3.10. Let $N(t)$ be a Poisson process with density v; then the process $(-1)^{N(t)}$ is one that jumps between two states, $+1$ and -1, with exponential waiting times between jumps.

(a) Find $E[(-1)^{N(t)}]$.

(b) By taking the limit as $t \to \infty$ in part (a), show that $P[(-1)^{N(t)} = 1] \to \frac{1}{2}$.

3.11. Let $N(t)$ be a Poisson process with density v. Using the results of Problem 3.10, find

(a) $E\left[\int_0^T (-1)^{N(t)} \, dt \right]$.

(b) $\left[E\left[\int_0^T (-1)^{N(t)} \, dt \right] \right]^2$

What are the limits as $v \to \infty$ in parts (a) and (b)?

3.12. Use the recursion formula (3.1.6) to develop a linear system of equations of the form

$$\frac{d\mathbf{P}}{dt} = \mathbf{AP}$$

Find the fundamental solution to this system with the initial condition

$$\mathbf{P}(0) = (1, 0, \ldots, 0)^T$$

3.13. (a) Show that $\lim_{h \to 0} P[N(t + h) - N(t) > \epsilon] = 0$.

(b) Find $\lim_{h \to 0} E\left[\dfrac{N(t + h) - N(t)}{h} \right]$.

(c) Show that $\lim_{h \to 0} E\left[\left(\dfrac{N(t + h) - N(t)}{h} \right)^2 \right]$ is infinite.

3.14. Let $U_1, \ldots, U_n$ be independent random variables uniformly distributed on $(0, s]$, and $U_{(1)}, \ldots, U_{(n)}$ their order statistics. Find $P[U_{(i)} < \tau]$ by

(a) integration of the joint density (3.6.3).

(b) computation of a conditional probability.

3.15. Use a symmetry argument to give an alternative proof of the formula (3.6.3) for the joint density of the order statistics. What is the formula if X_k, $k = 1, \ldots, n$ are independent, each with density $f_X(x)$?

3.16. If X_k, $k = 1, \ldots, n$ are independent random variables, each uniform on $[0, 1]$, show that the density for the jth order statistic $X_{(j)}$ is

$$g(x) = ax^{j-1}(1 - x)^{n-j}$$

where a is a normalizing constant. This is called a beta distribution. What is the formula if the X_k's each have density $f_X(x)$?

3.17. If $X_1, \ldots, X_n$ are independent identically exponentially distributed random variables, and $X_{(1)}, \ldots, X_{(n)}$ are their order statistics, show that

$$Y_1 = nX_{(1)}, \qquad Y_k = (n + 1 - k)(X_{(k)} - X_{(k-1)})$$

are independent with the same distribution as the X_i's.

3.18. If $X_k, k = 1, 2, \ldots$ are independent identically distributed, $S_0 = 0, S_n = X_1 + \cdots + X_n$, then $N(t) = \max \{n | S_n \leq t\}$ is called a renewal process. If X_k are exponential, we have shown that $N(t)$ is the Poisson process. What is the distribution of $N(t)$ if the X_k's are binomial?

3.19. Consider the random sum $S_N(t) = X_1 + \cdots + X_N$, where the X_k's are independent identical binomial random variables with $P[X_k = 1] = t$, and N is Poisson and independent of the X_k's. Show that $S_N(t)$ is a description of the Poisson process for $t \in [0, 1]$. Use the characteristic function of a binomial random variable to find the characteristic function of $S_N(t)$. Use the inversion formula to find the density of $S_N(t)$. This will involve the Dirac δ-function.

CHAPTER

4

Shot Noise

The events in a Poisson process are assumed to be instantaneous. Certainly this is a reasonable approximation for subatomic particle events. However, the instrument that records these events converts the arrival of a particle into a voltage impulse $f(t)$, where t is the time elapsed after the arrival of the particle. This function is called the effect function, or impulse response function, associated with the instrument. The voltage at any time t is then the sum of the various effect functions associated with the arrivals that have taken place up until time t. This phenomenon was first observed in audio circuits, where it was due to the arrivals of electrons at the anode of a vacuum tube and sounded like gunshots, so it was termed "shot noise." We will consider shot noise as the random process given by the sum of the effect functions, each produced by the occurrence of an event from a Poisson process.

4.1 SHOT NOISE AND ITS STATISTICAL MOMENTS

We consider first the sum of the effects due to the arrivals that take place in the finite time interval $(-T/2, T/2]$, and then later let $T \to \infty$ to get the effect from all emissions. Let $T_1, T_2, \ldots$ be independent identically distributed random variables uniform on $(-T/2, T/2]$ and N be a Poisson random variable independent of the T_i's with

$$P[N = k] = \frac{(vT)^k e^{-vT}}{k!}$$

129

where ν is the average number of emissions per second. Suppose that the arrival of a particle produces an effect $f(t)$, and that the effect functions of the various particles add linearly. The total effect at time t, due to particles arriving during $(-T/2, T/2]$ will then have the same distribution as the random variable

$$\sum_{j=1}^{N} f(t - T_j)$$

To find the first and second moments, we use its characteristic function

$$E\left[\exp\left(i\xi \sum_{j=1}^{N} f(t - T_j) \right) \right] \tag{4.1.1}$$

taking expectation on N

$$= \sum_{k=0}^{\infty} E\left[\exp\left(i\xi \sum_{j=1}^{k} f(t - T_j) \right) \right] P[N = k]$$

by independence of the T_j's

$$= \sum_{k=0}^{\infty} \left(\prod_{j=1}^{k} E[\exp (i\xi f(t - T_j))] \right) P[N = k]$$

since the T_j's are identically distributed

$$= \sum_{k=0}^{\infty} (E[\exp (i\xi f(t - T_1))])^k P[N = k]$$

or

$$= \sum_{k=0}^{\infty} (E[e^{i\xi f(t - T_1)}])^k \frac{(\nu T)^k e^{-\nu T}}{k!}$$

and writing the infinite series as an exponential function,

$$= \exp (-\nu T) \exp (\nu T E[e^{i\xi f(t - T_1)}])$$

Since T_1 is uniform on $(-T/2, T/2]$,

$$E[e^{i\xi f(t - T_1)}] = \int_{-T/2}^{T/2} e^{i\xi f(t - x)} \frac{1}{T} dx$$

and substituting this into (4.1.1) gives

$$E[\exp\left(i\xi \sum_{j=1}^{N} f(t - T_j)\right)] = \exp(-vT) \exp\left(vT \int_{-T/2}^{T/2} e^{i\xi f(t-x)} \frac{1}{T} dx\right)$$

$$= \exp\left[v \int_{-T/2}^{T/2} (e^{i\xi f(t-x)} - 1) dx\right]$$

(4.1.2)

By (1.7.8) we conclude from the first derivative of (4.1.2) that

$$E\left[\sum_{j=1}^{N} f(t - T_j)\right] = v \int_{-T/2}^{T/2} f(t - x) dx$$

(4.1.3)

and from the second derivative

$$E\left[\left(\sum_{j=1}^{N} f(t - T_j)\right)^2\right] = \left[v \int_{-T/2}^{T/2} f(t - x) dx\right]^2 + v \int_{-T/2}^{T/2} f^2(t - x) dx.$$

(4.1.4)

Since

$$\lim_{T\to\infty} \sum_{j=1}^{N} f(t - T_j)$$

gives the effect from all emissions, we can find the first and second moments of the shot noise process by taking the limit as $T \to \infty$ in (4.1.3) and (4.1.4). Assuming that $f(t) = 0$ for $t < 0$, this gives

$$\lim_{T\to\infty} E\left[\sum_{j=1}^{N} f(t - T_j)\right] = v \int_{0}^{\infty} f(u) du$$

(4.1.5)

and

$$\lim_{T\to\infty} E\left[\left(\sum_{j=1}^{N} f(t - T_j)\right)^2\right] = \left[v \int_{0}^{\infty} f(u) du\right]^2 + v \int_{0}^{\infty} f^2(u) du$$

(4.16)

These results are sometimes referred to as Campbell's theorem. Note that these are independent of t.

Linear systems

We can put our discussion of shot noise into a more general setting, by mathematically modeling the sequence of particle arrivals as a sum of δ-functions

$$X(t) = \sum_{i=1}^{\infty} \delta(t - S_i)$$

where S_i is the time of the ith occurrence of an event from a Poisson process. The output from the measuring device is then the linear combination of the outputs due to each arrival:

$$Y(t) = \sum_{i=1}^{\infty} f(t - S_i)$$

Systems for which a linear combination of inputs result in a similar combination of the corresponding outputs are called **linear filters** or **linear systems.** In other words, L is a linear system if

$$L(a_1 X_1(t) + a_2 X_2(t)) = a_1 L(X_1(t)) + a_2 L(X_2(t))$$

This implies under general conditions that

$$L\left(\int_{-\infty}^{\infty} a(s)X(t)\, ds \right) = \int_{-\infty}^{\infty} a(s)L(X(t))\, ds$$

The function

$$L(\delta(t)) = f(t)$$

is called the **effect function,** or **impulse response function,** of the system. A linear system is said to be **time invariant** if

$$L(X(t)) = Y(t)$$

implies, for $W(t) = X(t + s)$,

$$L(W(t)) = Y(t + s)$$

For a general input $X(t)$ the output of a time invariant linear system is given by the convolution

$$L(X(t)) = \int_{-\infty}^{\infty} X(t - s)f(s)\, ds$$

To see this write $X(t)$ as

$$X(t) = \int_{-\infty}^{\infty} X(s)\, \delta(t - s)\, ds$$

so that

$$L(X(t)) = L\left(\int_{-\infty}^{\infty} X(s)\, \delta(t - s)\, ds\right) = \int_{-\infty}^{\infty} X(s)L(\delta(t - s))\, ds$$

and by time invariance

$$= \int_{-\infty}^{\infty} X(s)f(t - s)\, ds = \int_{-\infty}^{\infty} X(t - s)f(s)\, ds$$

The expectation of the output is given by

$$E[L(X(t))] = \int_{-\infty}^{\infty} E[X(t - s)]f(s)\, ds = L(E[X(t)])$$

This formula allows us to express moments of the output in terms of the moments of the input, that is,

$$E[L_1(X(t_1))L_2(X(t_2))] = L_1(E[X(t_1)L_2(X(t_2))]) = L_1(L_2(E[X(t_1)X(t_2)]))$$

where the subscript on L_i indicates which of the t_i's is the dummy variable for each L_i. For example, the second moment of the output is given calculating the above and then setting $t_1 = t_2 = t$.

Shot noise is the output of a linear filter where the input is a sum of delta functions distributed according to a Poisson process. Such a sum is, formally, the derivative of a Poisson process. Suppose that $X(t)$ is shot noise due to arrivals in $(0, \infty)$, and $N(t)$ the corresponding Poisson process. We find the first and second moments of shot noise by finding $E[X(t)]$ and $E[X^2(t)]$ as $t \to \infty$. Let D be the linear system of differentiation with respect to t, and M be the linear system with impulse response function $f(t)$. The composition of linear filters is a linear filter. Thus

$$E[X(t)] = E[M(D(N(t)))]$$
$$= M(D(E[N(t)]))$$
$$= M(D(vt))$$
$$= M(v)$$
$$= \int_{-\infty}^{t} vf(u)\, du$$

Letting $t \to \infty$ gives the first moment.

To find the second moment, we first need to find

$$E[N(t + \Delta t)N(t)] = E[N^2(t) + N(t)N_\Delta]$$
$$= E[N^2(t)] + E[N(t)N_\Delta]$$

which by independence

$$= E[N^2(t)] + E[N(t)]E[N_\Delta]$$
$$= vt + v^2t^2 + v^2t\,\Delta t$$
$$= vt + v^2t(t + \Delta t)$$

so that

$$E[N(t)N(s)] = v \min (t, s) + v^2st$$

Set the composition of M and D equal to L. The mixed second moment for $X(t)$ is then

$$E[X(t_1)X(t_2)] = L_1(L_2(E[N(t_1)N(t_2)]))$$
$$= L_1(L_2(v \min (t_1, t_2) + v^2t_1t_2))$$
$$= L_1(M_2(vH(t_1 - t_2) + v^2t_1))$$

where H is the Heaviside step function. Now,

$$E[X(t_1)X(t_2)] = L_1\left(\int_{-\infty}^{t_2} vH(t_1 - (t_2 - s))f(s)\,ds + \int_{-\infty}^{t_2} v^2t_1 f(s)\,ds\right)$$

and assuming $f(s) = 0$ for $s < 0$,

$$= L_1\left(v\int_{t_2-t_1}^{t_2} f(s)\,ds + v^2\int_0^{t_2} t_1 f(s)\,ds\right)$$

This is

$$E[X(t_1)X(t_2)] = M_1\left(vf(t_2 - t_1) + v^2\int_0^{t_2} f(s)\,ds\right)$$

or

$$= v\int_0^{t_1} f(t_2 - (t_1 - s))f(s)\,ds + v^2\int_0^{t_2} f(s)\,ds \int_0^{t_1} f(s)\,ds$$

Thus when $t_1 = t_2 = t$ we get

$$E[X^2(t)] = v \int_0^t f^2(s) \, ds + \left(v \int_0^t f^2(s) \, ds \right)^2$$

Letting $t \to \infty$ gives the second moment.

4.2 A RELATED LIMITING DISTRIBUTION

In this section we discuss the limiting distribution of a random variable related to shot noise. In the next section we use the results obtained to examine the limiting distribution of the shot noise process. We consider the distribution of the following random variable which arises if shot noise is written as a Fourier series,

$$X_k = \sqrt{2/vT} \sum_{j=1}^{N} \cos \omega_k T_j \tag{4.2.1}$$

where T_j, N are as before, and $\omega_k = 2k\pi/T \neq 0$. The characteristic function of X_k is given by

$$E\left[\exp\left(i\xi \sqrt{\frac{2}{vT}} \sum_{j=1}^{N} \cos \omega_k T_j \right) \right] = \sum_{n=0}^{\infty} E\left[\exp\left(i\xi \sqrt{\frac{2}{vT}} \sum_{j=1}^{n} \cos \omega_k T_j \right) \right] P[N = n]$$

which since the T_j's are independent identically distributed

$$= \sum_{n=0}^{\infty} \left(E\left[\exp\left(i\xi \sqrt{\frac{2}{vT}} \cos \omega_k T_1 \right) \right] \right)^n P[N = n]$$

inserting the distribution of N

$$= \sum_{n=0}^{\infty} \left(E\left[\exp\left(i\xi \sqrt{\frac{2}{vT}} \cos \omega_k T_1 \right) \right] \right)^n \frac{(vT)^n e^{-vT}}{n!}$$

$$= e^{-vT} \sum_{n=0}^{\infty} \frac{1}{n!} \left(vTE\left[\exp\left(i\xi \sqrt{\frac{2}{vT}} \cos \omega_k T_1 \right) \right] \right)^n$$

or

$$= \exp\left(-vT + vTE\left[\exp\left(i\xi \sqrt{\frac{2}{vT}} \cos \omega_k T_1 \right) \right] \right)$$

To find the limit of this as $\nu \to \infty$, we expand the exponent as follows:

$$-\nu T + \nu TE\left[\exp\left(i\xi\sqrt{\frac{2}{\nu T}}\cos \omega_k T_1\right)\right]$$

$$= -\nu T + \nu T\int_{-T/2}^{T/2} \exp\left(i\xi\sqrt{\frac{2}{\nu T}}\cos \omega_k t\right)\frac{1}{T}\,dt$$

$$= -\nu T + \nu \int_{-T/2}^{T/2}\left[1 + i\xi\sqrt{\frac{2}{\nu T}}\cos \omega_k t + \frac{(i\xi\sqrt{2/\nu T}\cos \omega_k t)^2}{2} + \cdots\right]dt$$

$$= -\frac{\xi^2}{T}\int_{-T/2}^{T/2} \cos^2\omega_k t\,dt + \text{terms multiplied by powers of } \nu^{-1/2}$$

$$= -\frac{1}{2}\xi^2 + \text{terms multiplied by powers of } \nu^{-1/2}$$

Therefore, we obtain

$$\lim_{\nu \to \infty}\left(-\nu T + \nu TE\left[\exp\left(i\xi\sqrt{\frac{2}{\nu T}}\cos \omega_k T_1\right)\right]\right) = -\tfrac{1}{2}\xi^2$$

and so

$$\lim_{\nu \to \infty} E\left[\exp\left(i\xi\sqrt{\frac{2}{\nu T}}\sum_{j=1}^{N}\cos \omega_k T_j\right)\right] = \exp\left(-\tfrac{1}{2}\xi^2\right) \qquad (4.2.2)$$

which is the characteristic function of a normal random variable of mean zero and variance 1, implying that the limiting distribution as $\nu \to \infty$ is normal of mean zero and variance 1.

Replacing $\cos \omega_k T_j$ by $\sin \omega_k T_j$ in the argument above gives the corresponding result that the limiting distribution of

$$Y_k = \sqrt{\frac{2}{\nu T}}\sum_{j=1}^{N}\sin \omega_k T_j \qquad (4.2.3)$$

is normal of mean zero and variance 1.

4.3 THE LIMITING DISTRIBUTION OF THE SHOT NOISE PROCESS

We now examine the distribution of the effect due to emissions in $(-T/2, T/2]$, as $\nu \to \infty$. We assume that the effect function is periodic, and express it as a Fourier series. Later in this chapter we find the limiting distribution for a general effect function by letting $T \to \infty$.

If f is periodic with period T, then by a change of variables (4.1.3) becomes

$$E\left[\sum_{j=1}^{N} f(t - T_j)\right] = v \int_{-T/2}^{T/2} f(x)\, dx$$

and similarly (4.1.4) gives the second moment, so the variance is

$$E\left[\left(\sum_{j=1}^{N} f(t - T_j)\right)^2\right] - \left[E\left[\sum_{j=1}^{N} f(t - T_j)\right]\right]^2 = v \int_{-T/2}^{T/2} f^2(x)\, dx$$

From these observations, it makes sense to investigate the limiting distribution, as $v \to \infty$, of

$$\sqrt{\frac{1}{v}}\left[\sum_{j=1}^{N} f(t - T_j) - v \int_{-T/2}^{T/2} f(x)\, dx\right]$$

which has mean zero and constant variance for all values of v. To do this, we use the Fourier expansion (2.2.4) and write

$$f(t - T_j) = \frac{a_0}{2} + \sum_{k=1}^{\infty} a_k \cos \omega_k(t - T_j) + b_k \sin \omega_k(t - T_j)$$

$$= \frac{a_0}{2} + \sum_{k=1}^{\infty} \left(\begin{array}{l} + (a_k \cos \omega_k T_j - b_k \sin \omega_k T_j) \cos \omega_k t \\ (a_k \sin \omega_k T_j + b_k \cos \omega_k T_j) \sin \omega_k t \end{array} \right)$$

Therefore,

$$\sqrt{\frac{1}{v}}\left[\sum_{j=1}^{N} f(t - T_j) - v \int_{-T/2}^{T/2} f(x\,(dx)\right]$$

$$= \sqrt{\frac{1}{v}}\left[\sum_{j=1}^{N}\left(\frac{a_0}{2} + \sum_{k=1}^{\infty}\left(\begin{array}{l} + (a_k \cos \omega_k T_j - b_k \sin \omega_k T_j) \cos \omega_k t \\ (a_k \sin \omega_k T_j + b_k \cos \omega_k T_j) \sin \omega_k t \end{array}\right)\right)\right.$$

$$\left. - v \int_{-T/2}^{T/2} f(x)\, dx\right]$$

or, setting

$$X_k = \sqrt{\frac{2}{vT}} \sum_{j=1}^{N} \cos \omega_k T_j$$

$$Y_k = \sqrt{\frac{2}{vT}} \sum_{j=1}^{N} \sin \omega_k T_j$$

(4.3.1)

and interchanging the order of summation

$$= \sqrt{\frac{T}{2}} \frac{a_0}{2} X_0 - \sqrt{v} \int_{-T/2}^{T/2} f(x) \, dx$$

$$+ \sqrt{\frac{T}{2}} \sum_{k=1}^{\infty} (a_k X_k - b_k Y_k) \cos \omega_k t + (a_k Y_k + b_k X_k) \sin \omega_k t$$

Since $X_0 = \sqrt{2/vT} \, N$ and

$$a_0 = \frac{2}{T} \int_{-T/2}^{T/2} f(x) \, dx$$

this is

$$\sqrt{\frac{1}{v}} \left[\sum_{j=1}^{N} f(t - T_j) - v \int_{-T/2}^{T/2} f(x) \, dx \right] = \sqrt{T} \frac{a_0}{2} \frac{N - vT}{\sqrt{vT}}$$

$$+ \sqrt{\frac{T}{2}} \sum_{k=1}^{\infty} (a_k X_k - b_k Y_k) \cos \omega_k t + (a_k Y_k + b_k X_k) \sin \omega_k t \quad (4.3.2)$$

As $v \to \infty$, it follows from the exercises at the end of this chapter that $(N - vT)/\sqrt{vT}$, X_k and Y_k become independent Gaussian random variables with mean zero and variance 1, so that the limiting distribution of (4.3.2) is the same as the distribution of

$$\sqrt{T} \frac{a_0}{2} Z + \sqrt{\frac{T}{2}} \sum_{k=1}^{\infty} (a_k X_k' - b_k Y_k') \cos \omega_k t + (a_k Y_k' + b_k X_k') \sin \omega_k t \quad (4.3.3)$$

where Z, X_k', and Y_k' are independent normal random variables of mean zero and variance 1. The exercises also show that the linear combinations of these independent normal random variables

$$P_k = a_k X_k' - b_k Y_k'$$

$$Q_k = a_k Y_k' + b_k X_k', \qquad k > 0$$

are also independent normal random variables with

$$E[P_k] = E[Q_k] = 0$$

$$\text{var } P_k = \text{var } Q_k = a_k^2 + b_k^2$$

By defining the normalized P_k and Q_k,

$$U_k = \frac{P_k}{\sqrt{\text{var } P_k}}$$

$$V_k = \frac{Q_k}{\sqrt{\text{var } Q_k}}$$

$$\quad (4.3.4)$$

we may finally conclude that

$$\sqrt{\frac{1}{\nu}}\left[\sum_{j=1}^{N} f(t - T_j) - \nu \int_{-T/2}^{T/2} f(x)\, dx\right]$$

has as its limiting distribution the distribution of

$$\sqrt{T}\frac{a_0}{2} Z + \sqrt{\frac{T}{2}} \sum_{k=1}^{\infty} (a_k^2 + b_k^2)^{1/2}(U_k \cos \omega_k t + V_k \sin \omega_k t) \qquad (4.3.5)$$

where Z, U_k and V_k are independent normal random variables of mean zero and variance 1. When expressed in terms of the complex Fourier series, noting that

$$(a_k^2 + b_k^2)^{1/2} = 2|\alpha(\omega_k)|, \qquad \frac{a_0}{2} = \alpha(0)$$

this has the form

$$\sqrt{T}\,\alpha(0)Z + \sqrt{2T} \sum_{k=1}^{\infty} |\alpha(\omega_k)|(U_k \cos \omega_k t + V_k \sin \omega_k t) \qquad (4.3.6)$$

4.4 RANDOM NOISE

In this section we discuss some properties of the random noise process given by (4.3.6),

$$W(t) = \sqrt{T}\,\alpha(0)Z + \sqrt{2T} \sum_{k=1}^{\infty} |\alpha(\omega_k)|(U_k \cos \omega_k t + V_k \sin \omega_k t) \qquad (4.4.1)$$

The joint characteristic function of $W(t)$ and $W(s)$ is

$$E[\exp (i(\xi W(t) + \eta W(s)))] = E\left[\left(\exp (i(\xi + \eta)\sqrt{T}\,\alpha(0)Z)\right.\right.$$
$$\times \prod_{k=1}^{\infty} \exp (i\sqrt{2T}\,|\alpha(\omega_k)|(U_k(\xi \cos \omega_k t + \eta \cos \omega_k s)$$
$$\left.\left. + V_k(\xi \sin \omega_k t + \eta \sin \omega_k s)))\right)\right]$$

$$(4.4.2)$$

which since Z, U_k, and V_k are normal, mean zero, variance 1 independent random variables

$$= \exp\left[-\frac{T}{2}(\xi + \eta)^2\alpha(0)^2\right]$$

$$\times \exp\left(-T\sum_{k=1}^{\infty}|\alpha(\omega_k)|^2(\xi\cos\omega_k t + \eta\cos\omega_k s)^2 + |\alpha(\omega_k)|^2(\xi\sin\omega_k t \right.$$

$$\left. + \eta\sin\omega_k s)^2\right)$$

$$= \exp\left(-\frac{T}{2}(\xi + \eta)^2\alpha(0)^2 - T\sum_{k=1}^{\infty}|\alpha(\omega_k)|^2(\xi^2 + \eta^2 + 2\eta\xi\cos\omega_k(t - s))\right)$$

$$= \exp\left(-\frac{T}{2}\sum_{k=-\infty}^{\infty}|\alpha(\omega_k)|^2(\xi^2 + \eta^2 + 2\eta\xi\cos\omega_k(t - s))\right)$$

This is the characteristic function of a bivariate normal distribution, so that the joint distribution of $W(t)$ and $W(s)$ is bivariate normal. With a little more effort, it can be shown that the higher dimensional joint distributions of the process are multivariate normal. Processes with this property are called Gaussian processes. We discuss these processes in greater detail in chapter 5.

In addition $W(t)$ is stationary, by which we mean that its joint distribution is unchanged under time translations. To show this, note that by using the substitution

$$U_k' = U_k\cos\omega_k s + V_k\sin\omega_k s$$

$$V_k' = V_k\cos\omega_k s - U_k\sin\omega_k s$$

then

$$W(t + s) = \sqrt{T}\,\alpha(0)Z + \sqrt{2T}\sum_{k=1}^{\infty}|\alpha(\omega_k)|(U_k\cos\omega_k(t + s) + V_k\sin\omega_k(t + s))$$

becomes

$$W(t + s) = \sqrt{T}\,\alpha(0)Z + \sqrt{2T}\sum_{k=1}^{\infty}|\alpha(\omega_k)|(U_k'\cos\omega_k t + V_k'\sin\omega_k t) \quad (4.4.3)$$

Now, observing that U_k', V_k', and Z are independent normal of mean zero and variance 1 gives the result.

A determining parameter for a Gaussian process $X(t)$ is its covariance $r(u, t)$ given by

$$r(u, t) = \text{cov} (X(u), X(t))$$
$$= E[(X(u) - \mu_u)(X(t) - \mu_t)]$$

For random noise this is

$$\text{cov} (W(u), W(t)) = E[W(u)W(t)] \tag{4.4.4}$$

which can be calculated explicitly,

$$= T\alpha^2(0) + 2T \sum_{k=1}^{\infty} |\alpha(\omega_k)|^2 (\cos \omega_k t \cos \omega_k u + \sin \omega_k t \sin \omega_k u)$$

$$= T \sum_{k=-\infty}^{\infty} |\alpha(\omega_k)|^2 \cos \omega_k (u - t)$$

This as expected from stationarity depends only on the difference of the two times. For stationary processes we shall write the covariance as a function of the time difference

$$r(\tau) = E[(X(t + \tau) - \mu_{t+\tau})(X(t) - \mu_t)] \tag{4.4.5}$$

where $\tau = u - t$, and for random noise we obtain

$$r(\tau) = T \sum_{k=-\infty}^{\infty} |\alpha(\omega_k)|^2 \cos \omega_k \tau \tag{4.4.6}$$

4.5   THE SPECTRAL DENSITY OF RANDOM NOISE

Here we consider what happens to random noise as $T \to \infty$. The formula for the covariance function then becomes an integral. Substituting (2.7.1) into (4.4.6) gives

$$r(\tau) = T \sum_{k=-\infty}^{\infty} \left| \frac{1}{T} \int_{-T/2}^{T/2} f(u)e^{i\omega_k u} \, du \right|^2 \cos \omega_k \tau$$

or

$$= \sum_{k=-\infty}^{\infty} \frac{2\pi}{T} \left| \frac{1}{2\pi} \int_{-T/2}^{T/2} f(u)e^{i\omega_k u} \, du \right|^2 \cos \omega_k \tau$$

Letting $T \to \infty$, and setting

$$A(\omega) = \int_{-\infty}^{\infty} f(u)e^{i\omega u} \, du \tag{4.5.1}$$

this becomes

$$r(\tau) = \frac{1}{2\pi} \int_{-\infty}^{\infty} |A(\omega)|^2 \cos \omega\tau \, d\omega$$

$$= \frac{1}{\pi} \int_{0}^{\infty} |A(\omega)|^2 \cos \omega\tau \, d\omega$$

(4.5.2)

The process resulting from letting $T \to \infty$ formally is the limit of (4.4.1) and can be expressed as a kind of integral, which is called a stochastic integral. The quantity $|A(\omega)|^2$ is called the spectral density of the process. The complete theory of stochastic integrals and differentials lies outside the scope of this book, but we perform some integrations involving stochastic processes in chapter 7. In our case one can check, by taking the limit as $T \to \infty$ in (4.4.2), that the resulting process is Gaussian and by considering (4.4.3) it is stationary. Further, noting that $|A(\omega)|^2$ is even,

$$r(\tau) = \frac{1}{2\pi} \int_{-\infty}^{\infty} |A(\omega)|^2 \cos \omega\tau \, d\omega$$

$$= \frac{1}{2\pi} \int_{-\infty}^{\infty} |A(\omega)|^2 e^{-i\omega t} \, d\omega$$

(4.5.3)

we see that $r(\tau)$ and the spectral density $|A(\omega)|^2$ are related by the Fourier inversion formula.

PROBLEMS

4.1. Suppose that X_i are independent identically distributed random variables, and that N is positive integer-valued random variable independent of the X_i's.

(a) Show that the characteristic function of

$$Z = \sum_{i=1}^{N} X_i$$

is of the form

$$C_Z(\xi) = E[(C_{X_1}(\xi))^N]$$

by using the property of conditional expectation

$$E[E[X|Y]] = E[X]$$

(b) Suppose that X_i are Bernoulli random variables, with

$$P[X_i = 1] = p$$

Use part (a) to show that Z has a Poisson distribution.

4.2. **(a)** Perform the differentiations of the characteristic function to verify formulas (4.1.4) and (4.1.5).

(b) Let $T \to \infty$ in these two formulas, and assuming that $f(t) = 0$ for $t < 0$, show (4.1.5) and (4.1.6). Why is this assumption natural?

4.3. Consider the random variable given in Section 4.1,

$$Z(t) = \sum_{j=1}^{N} f(t - T_j)$$

(a) Find its moment generating function.

(b) Find the variance of $Z(t)$.

(c) Use part (a) to show that in the limit as $T \to \infty$, the nth cumulant of $Z(t)$ is

$$\nu \int_0^\infty f^n(u)\, du$$

provided that $f(t) = 0$ for $t < 0$.

4.4. Consider the transformation $L(X(t)) = Y(t)$ given by

$$Y(t) = \int_{t-a}^{t} X(s)\, ds$$

(a) Show that this is a linear system.

(b) Find the impulse response function $f(t)$.

(c) Verify that L is given by

$$L(X(t)) = \int_{-\infty}^{\infty} X(t - s) f(s)\, ds$$

(d) Suppose that $r(t, s) = E[X(t)X(s)]$. Find $E[Y(t)Y(s)]$.

(e) Suppose that $X(t)$ is a sum of Dirac delta functions distributed according to a Poisson process $N(t)$ of parameter ν, given formally by

$$X(t) = \frac{dN(t)}{dt}$$

Suppose that $X(t)$ corresponds to the arrivals of particles at a counter. Describe $Y(t)$ in words.

(f) Under the conditions of part (e), find the mean and variance of $Y(t)$.

4.5. Consider the scaled Poisson process $Y(t)$:

$$Y(t) = \frac{1}{\sqrt{\nu}} N(t)$$

where $N(t)$ is a Poisson process of parameter ν. Show that it has covariance function

$$r(t, s) = E[(Y(t) - E[Y(t)])(Y(s) - E[Y(s)])]$$

which, as $\nu \to \infty$, becomes $r(t, s) = \min(t, s)$.

4.6. Show that the limiting distribution of

$$Y_k = \sqrt{2/\nu T} \sum_{j=1}^{N} \sin \omega_k T_j$$

is normal with mean zero and variance 1.

4.7. Calculate the limiting joint characteristic function of the random variables X_k and Y_k, $k \neq 0$ of (4.2.1) and (4.2.3), to show that in the limit they become independent normal random variables.

4.8. Set $k = 0$ in equation (4.2.1). What goes wrong with the proof that in the limit X_0 is normal of mean zero and variance 1?

4.9. We know from Problem 3.9 that in the limit as $\nu \to \infty$,

$$Z = \frac{N - \nu T}{\sqrt{\nu T}}$$

becomes normal of mean zero and variance 1. Show that in the limit Z becomes independent of both X_k and Y_k, $k \neq 0$.

4.10. Perform the change of variables necessary to get (4.3.5) from (4.3.3).

4.11. Using the characteristic function, show that if X and Y are independent normal random variables, $aX + bY$ is independent of $cX + dY$ if and only if $ac + bd = 0$. Show in general that if $X_1, X_2, \ldots$ are independent normal random variables and $\mathbf{X} = (X_1, \ldots, X_n)$, then $\sum_{k=1}^{n} a_k X_k$ is independent of $\sum_{k=1}^{n} b_k X_k$ if and only if $\sum_{k=1}^{n} a_k b_k = 0$.

4.12. As will be shown in Chapter 5, the characteristic function of a multivariate normal distribution is given by

$$C_\mathbf{X}(\boldsymbol{\xi}) = K \exp\left(-\tfrac{1}{2} \boldsymbol{\xi}^T \mathbf{R} \boldsymbol{\xi}\right)$$

where $\mathbf{R}$ is a positive definite matrix, $\mathbf{X} = (X_1, \ldots, X_n)$, $\boldsymbol{\xi} = (\xi_1, \ldots, \xi_n)^T$ and K is a constant depending on the matrix $\mathbf{R}$.

(a) Verify that (4.4.1) has a joint normal distribution, by showing that its characteristic function is of this form.

(b) In the same way, show that equation (4.4.2) is the characteristic function of a mean zero bivariate normal random variable.

4.13. If the effect function of shot noise is of the form

$$f(t) = e^{-\alpha t}, t \geq 0$$

what is the form of the spectral density function $|A(\omega)|^2$ and what is the form of the covariance (4.5.3)?

4.14. Show that if the variance of a stationary process is one, then the spectral density is a probability density.

4.15. If the effect function of shot noise is of the form

$$f(t) = \begin{cases} 1, & t \in [0, 1] \\ 0, & \text{else} \end{cases}$$

what is the form of the spectral density function $|A(\omega)|^2$ and what is the form of the covariance (4.5.3)?

4.16. Suppose that $Z(t) = a(X \cos \omega t + Y \sin \omega t)$, where X and Y are independent normal random variables of mean zero and variance 1.

(a) Find the mean function $\mu(t) = E[Z(t)]$.

(b) Find the covariance function $r(t, s) = E[(Z(t) - \mu(t))(Z(s) - \mu(s))]$.

(c) Show that the covariance function has the property

$$r(t, s) = r(t + \tau, s + \tau)$$

4.17. (a) Show that if $(E[XY])^2 = E[X^2]E[Y^2]$, then $X = cY$ almost surely, for some constant c. This is a special case of the Cauchy–Schwarz inequality.

(b) Suppose the covariance function of a mean zero stochastic process $Y(t)$ is constant

$$r(s, t) = E[Y(t)Y(s)] = K$$

By applying part (a), show that $Y(t)$ is deterministic, and in particular satisfies

$$Y(t) = h(t)Y(0)$$

for some function $h(t)$.

(c) Show that a general process $Y(t)$ with

$$E[(Y(t) - E[Y(t)])(Y(s) - E[Y(s)])] = K$$

has

$$Y(t) = h(t)Y(0) + g(t)$$

Express h and g in terms of the mean and variance of $Y(t)$.

4.18. Let

$$Z(t) = X \cos \omega t - Y \sin \omega t$$

where X and Y are independent normal random variables of mean zero and variance 1.

(a) Show that

$$R = \sqrt{X^2 + Y^2}$$

has the density

$$f_R(r) = \begin{cases} re^{-(1/2)r^2}, & 0 \le r < \infty \\ 0, & r \le 0 \end{cases}$$

(b) Suppose that R and θ are independent, where θ is uniform on $(0, 2\pi]$. Show $R \cos \theta$ and $R \sin \theta$ are independent and normally distributed with mean zero and variance 1.

(c) Conclude that

$$W(t) = R \cos (\omega t + \theta)$$

has the same joint distributions as $Z(t)$.

CHAPTER

5

Gaussian Processes

Recall that by definition a stochastic process $X(t)$ is a jointly distributed family of random variables, parametrized by t. Suppose we are given a stochastic process such that for each $n > 0$ there exists a nonnegative function $W(x_1, t_1; x_2, t_2; \ldots; x_n, t_n)$, continuous in the x_i's, and satisfying for all choices of a_i,

$$P[X(t_1) \leq a_1, X(t_2) \leq a_2, \ldots, X(t_n) \leq a_n]$$

$$= \int_{-\infty}^{a_1} \cdots \int_{-\infty}^{a_n} W(x_1, t_1; x_2, t_2; \ldots; x_n, t_n) \, dx_1 \, dx_2 \ldots dx_n \tag{5.0.1}$$

The functions W are then called the joint density functions of the process, and it follows immediately that

$$\int_{-\infty}^{\infty} \cdots \int_{-\infty}^{\infty} W(x_1, t_1; \ldots; x_n, t_n) \, dx_1 \ldots dx_n = 1 \tag{5.0.2}$$

The knowledge of the joint density functions for all values of x_i, t_i, and n defines the process completely. It can be shown that there are two necessary and sufficient conditions, called Kolmogorov's consistency requirements, for a family of functions W satisfying (5.0.2) to be the joint density function of a stochastic process:

(i) $\quad \int_{-\infty}^{\infty} W(x_1, t_1; \ldots; x_n, t_n) \, dx_i =$

$$\tag{5.0.3}$$

$$W(x_1, t_1; \ldots; x_{i-1}, t_{i-1}; x_{i+1}, t_{i+1}; \ldots; x_n, t_n)$$

147

(ii)   if σ is any permutation of the subscripts $1, \ldots, n$, then

$$W(x_{\sigma(1)}, t_{\sigma(1)}; \ldots; x_{\sigma(n)}, t_{\sigma(n)}) = W(x_1, t_1; \ldots; x_n, t_n) \qquad (5.0.4)$$

In other words, these two relations can be used to define a family of random variables $X(t)$ for which W is the joint density. We shall not prove this, since the proof requires a subtle measure-theoretic argument (the so-called Kolmogorov reconstruction theorem), but we use it implicitly when we describe stochastic processes based on their joint densities alone.

5.1   THE GAUSSIAN PROCESS

A stochastic process $X(t)$ is said to be Gaussian if its joint density function W is multivariate normal,

$$W(x_1, t_1; \ldots; x_n, t_n) = \frac{\sqrt{\det \mathbf{A}}}{\sqrt{(2\pi)^n}} \exp\left[-\tfrac{1}{2} \sum_{i=1}^{n} \sum_{j=1}^{n} a_{ij}(x_i - \mu_i)(x_j - \mu_j) \right]$$
$$(5.1.1)$$

where the matrix $\mathbf{A} = (a_{ij})$ is symmetric positive definite and depends on $t_1, t_2, \ldots, t_n$.

Let us explain why this is called the multivariate normal distribution. It is clear that any reasonable characterization of multivariate normal random variables should include those of the form

$$\mathbf{Y} = (Y_1, Y_2, \ldots, Y_n)^{\mathrm{T}}$$

where the Y_i's are independent normal random variables. One can show that every random variable with density (5.1.1) is essentially of this type. For convenience, we assume that each Y_i has mean zero and variance $1/d_i^2$, $d_i > 0$. The joint density function $f_{\mathbf{Y}}$ is, by independence, the product of the density functions of the individual Y_i's

$$f_{\mathbf{Y}}(y_1, \ldots, y_n) = \prod_{i=1}^{n} d_i \frac{1}{\sqrt{2\pi}} \exp\left[-\tfrac{1}{2}(d_i y_i)^2 \right]$$

which we can write

$$= \frac{\sqrt{\det \mathbf{D}}}{\sqrt{(2\pi)^n}} \exp\left(-\tfrac{1}{2} \mathbf{y}^{\mathrm{T}} \mathbf{D} \mathbf{y} \right)$$

where $\mathbf{D}$ is a diagonal matrix with d_i^2 as its ith diagonal element, and $\mathbf{y} = (y_1, \ldots, y_n)^\mathrm{T}$. We would like normality to be a property that is independent of the choice of coordinate system used to describe the random variable $\mathbf{Y}$. Since changes in coordinates correspond to orthogonal transformations, under any orthogonal transformation

$$\mathbf{X} = \mathbf{QY}$$

where $\mathbf{Q}$ is an orthogonal matrix, $\mathbf{X}$ should have a multivariate normal distribution. The density for $\mathbf{X}$, since $\mathbf{Q}^\mathrm{T}\mathbf{X} = \mathbf{Y}$, is

$$f_{\mathbf{X}} = \frac{\sqrt{\det \mathbf{D}}}{\sqrt{(2\pi)^n}} \exp\left[-\tfrac{1}{2}(\mathbf{Q}^\mathrm{T}\mathbf{x})^\mathrm{T}\mathbf{D}\mathbf{Q}^\mathrm{T}\mathbf{x}\right] |\det \mathbf{Q}|$$

or, setting $\mathbf{A} = \mathbf{QDQ}^\mathrm{T}$, and noting that $|\det \mathbf{Q}| = 1$, we have

$$f_{\mathbf{X}} = \frac{\sqrt{\det \mathbf{A}}}{\sqrt{(2\pi)^n}} \exp\left[-\tfrac{1}{2}(\mathbf{x}^\mathrm{T}\mathbf{A}\mathbf{x})\right]$$

$$= \frac{\sqrt{\det \mathbf{A}}}{\sqrt{(2\pi)^n}} \exp\left(-\tfrac{1}{2}\sum_{i=1}^{n}\sum_{j=1}^{n} a_{ij}x_i x_j\right)$$

By a translation in $\mathbf{x}$ we obtain the general form (5.1.1). Since matrices of the form $\mathbf{A} = \mathbf{QDQ}^\mathrm{T}$, where $\mathbf{D}$ is diagonal with positive diagonal entries, are exactly the symmetric positive definite matrices, it is clear that every multivariate normal density arises in this fashion.

   We can make a further observation. Since every real symmetric matrix is diagonalizable, it follows that any probability density of the form

$$f(x_1, \ldots, x_n) = C \exp\left(-\tfrac{1}{2}\mathbf{x}^\mathrm{T}\mathbf{A}\mathbf{x}\right)$$

where C is a normalizing constant and $\mathbf{A}$ is assumed only symmetric, must have $\mathbf{A}$ positive definite. This is because diagonalizability implies, by a change of variables, that the normalization condition can be written as

$$\int_{-\infty}^{\infty} \cdots \int_{-\infty}^{\infty} K \exp\left(-\tfrac{1}{2}\sum_{i=1}^{n} \lambda_i x_i^2\right) dx_1 \cdots dx_n = 1$$

where the λ_i's are the eigenvalues of $\mathbf{A}$ and K is the product of C with the Jacobian of the change of variables. This integral is finite only if all the λ_i's are positive.

5.2   CALCULATION OF THE CHARACTERISTIC FUNCTION

The description given in the preceding section can be exploited to find the characteristic function of $\mathbf{X} = \mathbf{QY}$. We first calculate

$$C_\mathbf{Y}(\boldsymbol{\eta}) = E[\exp(i\boldsymbol{\eta}^T\mathbf{Y})] \tag{5.2.1}$$

by independence

$$= \prod_{j=1}^n E[\exp(i\eta_j Y_j)]$$

and by (1.7.7)

$$= \prod_{j=1}^n \exp\left(-\frac{1}{2}\frac{\eta_j^2}{d_j^2}\right)$$

$$= \exp\left(-\frac{1}{2}\boldsymbol{\eta}^T\mathbf{D}^{-1}\boldsymbol{\eta}\right)$$

Now we can write

$$C_\mathbf{X}(\boldsymbol{\xi}) = E[\exp(i\boldsymbol{\xi}^T\mathbf{X})]$$
$$= E[\exp(i\boldsymbol{\xi}^T\mathbf{QY})]$$

by (5.2.1) with $\boldsymbol{\eta}^T = \boldsymbol{\xi}^T\mathbf{Q}$

$$= \exp[-\tfrac{1}{2}(\mathbf{Q}^T\boldsymbol{\xi})^T\mathbf{D}^{-1}(\mathbf{Q}^T\boldsymbol{\xi})]$$
$$= \exp[-\tfrac{1}{2}(\boldsymbol{\xi}^T\mathbf{QD}^{-1}\mathbf{Q}^T\boldsymbol{\xi})]$$
$$= \exp(-\tfrac{1}{2}\boldsymbol{\xi}^T\mathbf{A}^{-1}\boldsymbol{\xi})$$

Defining $\mathbf{R} = \mathbf{A}^{-1}$, this takes the form

$$C_\mathbf{X}(\boldsymbol{\xi}) = e^{-(1/2)\boldsymbol{\xi}^T\mathbf{R}\boldsymbol{\xi}} \tag{5.2.2}$$

or denoting the entries of $\mathbf{R}$ as (r_{ij})

$$= \exp\left(-\frac{1}{2}\sum_{i=1}^n\sum_{j=1}^n \xi_i\xi_j r_{ij}\right)$$

This implies, by a change of variables, that the characteristic function of a Gaussian process $X(t)$ with joint density (5.1.1) is

$$C_{\mathbf{X}}(\xi) = \exp\left[-\tfrac{1}{2} \sum_{i=1}^{n} \sum_{j=1}^{n} \xi_i \xi_j r_{ij} + i \sum_{i=1}^{n} \xi_i \mu_i \right] \tag{5.2.3}$$

where $\mathbf{X} = (X(t_1), \ldots X(t_n))^T$ and $(r_{ij}) = \mathbf{A}^{-1}$.

5.3 THE COVARIANCE MATRIX

In this section we show that the covariances of a Gaussian process $X(t)$ are the entries of the matrix $\mathbf{R} = \mathbf{A}^{-1}$, that is,

$$E[(X(t_p) - \mu_p)(X(t_q) - \mu_q)] = r_{pq}$$

For convenience we shall assume that $\mu_i = 0$. Note by (1.8.7) that if $C_{\mathbf{X}}$ is the joint characteristic function of a process $X(t)$, then

$$\frac{d}{d\xi_q}[C_{\mathbf{X}}(\xi)]\bigg|_{\xi_1 = \xi_2 = \cdots = \xi_n = 0} = iE[X(t_q)] \tag{5.3.1}$$

and

$$\frac{d^2}{d\xi_p \, d\xi_q}[C_{\mathbf{X}}(\xi)]\bigg|_{\xi_1 = \xi_2 = \cdots = \xi_n = 0} = -E[X(t_q)X(t_p)] \tag{5.3.2}$$

For a mean zero Gaussian process by (5.2.2) we have

$$\frac{d}{d\xi_q}[C_{\mathbf{X}}(\xi)] = -\frac{1}{2}\left[\sum_{j=1}^{n} (r_{qj} + r_{jq})\xi_j \right] \exp\left(-\tfrac{1}{2} \sum_{i=1}^{n} \sum_{j=1}^{n} \xi_i \xi_j r_{ij} \right)$$

or since $\mathbf{R}$ is symmetric

$$= -\left(\sum_{j=1}^{n} r_{qj}\xi_j \right) \exp\left(-\tfrac{1}{2} \sum_{i=1}^{n} \sum_{j=1}^{n} \xi_i \xi_j r_{ij} \right)$$

Substituting this into (5.3.1) gives the expected result

$$E[X(t_q)] = 0$$

The second derivative is

$$\frac{d}{d\xi_p d\xi_q}[C_{\mathbf{X}}(\xi)] = -r_{qp}\left[\exp\left(-\tfrac{1}{2}\sum_{i=1}^{n}\sum_{j=1}^{n}\xi_i\xi_j r_{ij}\right)\right]$$

$$-\sum_{j=1}^{n}r_{qj}\xi_j\left[-\sum_{j=1}^{n}r_{pj}\xi_j\exp\left(-\tfrac{1}{2}\sum_{i=1}^{n}\sum_{j=1}^{n}\xi_i\xi_j r_{ij}\right)\right]$$

which by (5.3.2) yields

$$E[X(t_q)X(t_p)] = r_{qp} \tag{5.3.3}$$

Since $(r_{pq}) = \mathbf{A}^{-1}$, this implies that the joint density

$$W = \frac{\sqrt{\det \mathbf{A}}}{\sqrt{(2\pi)^n}}e^{-(1/2)\mathbf{x}^T\mathbf{A}\mathbf{x}}$$

of a mean zero Gaussian process is completely determined by its covariances. The density of a general Gaussian process is completely determined by its means and covariances.

5.4 A RULE FOR CALCULATING HIGHER CORRELATIONS

The method of the preceding section can be used to find higher correlations. That is, by further differentiating the characteristic function and then setting $\xi_1 = \cdots = \xi_n = 0$ one can, in principle, calculate any mixed moment $E[X(t_1)X(t_2)\cdots X(t_n)]$. We shall see that for a mean zero Gaussian process the following rule results from such calculations:

$$E[X(t_1)X(t_2)\cdots X(t_n)] = \begin{cases} 0, & n \text{ odd} \\ \sum E[X(t_i)X(t_j)]\cdots E[X(t_m)X(t_k)], & n \text{ even} \end{cases}$$

where the sum is taken over all decompositions into completely disjoint pairs of time points. Thus for $n = 4$ we have

$$\begin{aligned} E[X(t_1)X(t_2)X(t_3)X(t_4)] &= E[X(t_1)X(t_2)]E[X(t_3)X(t_4)] \\ &+ E[X(t_1)X(t_3)]E[X(t_2)X(t_4)] \\ &+ E[X(t_1)X(t_4)]E[X(t_3)X(t_2)] \end{aligned} \tag{5.4.1}$$

The proof of this special case is given below, and the general formula is left as an exercise.

Introduce the notation

$$C = C_X(\xi) = \exp\left(-\tfrac{1}{2}\sum_{i=1}^{4}\sum_{j=1}^{4}\xi_i\xi_j r_{ij}\right)$$

and

$$L_p = \sum_{j=1}^{4} r_{pj}\xi_j$$

then by (5.3.3),

$$\frac{dC}{d\xi_p} = -L_p C$$

and

$$\frac{dL_p}{d\xi_q} = r_{pq}$$

Computing

$$\frac{d^4 C}{d\xi_1\, d\xi_2\, d\xi_3\, d\xi_4} = \frac{d^3(-L_4 C)}{d\xi_1\, d\xi_2\, d\xi_3}$$

$$= \frac{d^2(-r_{43}C + L_4 L_3 C)}{d\xi_1\, d\xi_2}$$

$$= \frac{d(r_{43}L_2 C + r_{42}L_3 C + r_{32}L_4 C - L_4 L_3 L_2 C)}{d\xi_1}$$

$$= r_{43}r_{21}C - r_{43}L_2 L_1 C + r_{42}r_{31}C - r_{42}L_3 L_1 C$$

$$\quad + r_{32}r_{41}C - r_{32}L_4 L_1 - r_{41}L_3 L_2 C - r_{31}L_4 L_2 C$$

$$\quad - r_{21}L_4 L_3 C + L_4 L_3 L_2 L_1 C$$

and setting $\xi_1 = \xi_2 = \xi_3 = \xi_4 = 0$ gives

$$E[X(t_1)X(t_2)X(t_3)X(t_4)] = r_{43}r_{21} + r_{42}r_{31} + r_{32}r_{41}$$

which is (5.4.1). A close examination of the pattern of the subscripts and signs yields the general formula. We note that the t_i's do not have to be distinct numbers; for instance, one can use this to calculate the well-known formula, for Gaussian processes,

$$E[X^4(t)] = 3(E[X^2(t)])^2$$

It can be shown that if $X(t)$ is a mean zero process satisfying the rule for higher correlations then $X(t)$ is a Gaussian process. This follows from the fact that under general hypotheses, the moments of a random variable completely determine its distribution. Since the rule for higher correlations determines the moments of $(X(t_1), \ldots, X(t_n))$ to be those of a multivariate normal distribution, the joint distribution function must therefore be normal, so $X(t)$ will be Gaussian. Of course, if $X(t)$ is not a mean zero process, it will be Gaussian if $Z(t) = X(t) - \mu_t$ satisfies the rule for higher correlations.

5.5   THE INTEGRAL OF A GAUSSIAN PROCESS

In this section we show that the integral of a Gaussian process is itself a Gaussian process. We begin by using some of the methods of this chapter to give a few results on random variables that have a joint normal distribution. For convenience we shall assume the random variables have mean zero, but the results can be trivially extended to random variables with nonzero means.

If $X_1, X_2, \ldots, X_n$, are mean zero random variables with normal joint density, then for any real numbers $b_1, \ldots, b_n$,

$$Y = \sum_{k=1}^{n} b_k X_k$$

is a normal random variable. We can show this by finding the characteristic function of Y,

$$C_Y(\eta) = E[e^{i\eta Y}] = E\left[\exp\left(i\eta \sum_{k=1}^{n} b_k X_k\right)\right]$$

or if $\eta b_k = \xi_k$,

$$= E\left[\exp\left(i \sum_{k=1}^{n} \xi_k X_k\right)\right]$$

which by (5.2.2) is

$$= \exp\left(-\tfrac{1}{2} \sum_{i=1}^{n} \sum_{j=1}^{n} \xi_i \xi_j r_{ij}\right)$$

or

$$= \exp\left(-\tfrac{1}{2}\eta^2 \sum_{i=1}^{n} \sum_{j=1}^{n} b_i b_j r_{ij}\right)$$

This is the characteristic function of a normal random variable with mean zero and variance

$$\sum_{i=1}^{n} \sum_{j=1}^{n} b_i b_j r_{ij}$$

The converse is also true. Given mean zero random variables $X_1, X_2, \ldots, X_n$, if for any real numbers $b_1, \ldots, b_n$,

$$Y = \sum_{k=1}^{n} b_k X_k$$

is a normal random variable, then $X_1, X_2, \ldots, X_n$ are jointly normally distributed. By setting

$$r_{ij} = E[X_i, X_j]$$

we know that

$$E\left[\left(\sum_{i=1}^{n} b_i X_i\right)^2\right] = \sum_{i=1}^{n} \sum_{j=1}^{n} b_i b_j r_{ij} \tag{5.5.1}$$

is the variance of the normal random variable Y, so that

$$C_Y(\eta) = e^{-(1/2)\eta^2 \mathrm{var} Y}$$

or

$$E\left[\exp\left(i\eta \sum_{k=1}^{n} b_k X_k\right)\right] = \exp\left(-\tfrac{1}{2}\eta^2 \sum_{i=1}^{n} \sum_{j=1}^{n} b_i b_j r_{ij}\right)$$

Setting $\xi_k = \eta b_k$, this is

$$E\left[\exp\left(i \sum_{k=1}^{n} \xi_k X_k\right)\right] = \exp\left(-\tfrac{1}{2} \sum_{i=1}^{n} \sum_{j=1}^{n} \xi_i \xi_j r_{ij}\right)$$

and noting (r_{ij}) is positive definite by (5.5.1) gives the result. As an application of this, we show the following.

If $X(t)$ is a mean zero Gaussian process whose covariance

$$r(s, t) = E[X(s)X(t)]$$

is a continuous function of s and t, then for any deterministic piecewise continuous function $a(t)$,

$$Z(t) = \int_0^t a(s)X(s) \, ds \qquad (5.5.2)$$

is a Gaussian process. To demonstrate this fact, we first show that for a fixed τ, $Z(\tau)$ is a normal random variable. This is done by calculating the characteristic function of the associated Riemann sum, and passing to a limit. Let

$$\Delta t = \frac{\tau}{n}$$

$$t_k = k \, \Delta t, \qquad k = 1, 2, \ldots, n$$

Since $X(t_k)$, $k = 1, 2, \ldots, n$ are jointly normally distributed, we know that

$$E\left[\exp\left(i \sum_{k=1}^{n} \xi_k X(t_k) \right) \right] = \exp\left[-\tfrac{1}{2} \sum_{k=1}^{n} \sum_{j=1}^{n} \xi_k \xi_j r(t_k, t_j) \right]$$

Setting

$$\xi_k = a(t_k)\xi \, \Delta t$$

we obtain

$$E\left[\exp\left(i\xi \sum_{k=1}^{n} a(t_k) X(t_k) \, \Delta t \right) \right] = \exp\left[-\tfrac{1}{2}\xi^2 \, \Delta t^2 \sum_{k=1}^{n} \sum_{j=1}^{n} a(t_k)a(t_j)r(t_k, t_j) \right]$$

Letting $\Delta t \to 0$ this becomes

$$E\left[\exp\left(i\xi \int_0^\tau a(t)X(t) \, dt \right) \right] = \exp\left[-\tfrac{1}{2}\xi^2 \int_0^\tau \int_0^\tau a(t)a(s)r(s, t) \, ds \, dt \right]$$

which says that Z is a normal random variable of mean zero and variance

$$\int_0^\tau \int_0^\tau a(t)a(s)r(s, t) \, ds \, dt$$

We may now conclude that the integral (5.5.2) in addition to being normally distributed, is in fact a Gaussian process. By the first part of this section, it suffices to check that every linear combination of the form

$$b_1 Z(t_1) + \cdots b_n Z_n(t_n) = b_1 \int_0^{t_1} a(s)V(s) \, ds + \cdots + b_n \int_0^{t_n} a(s)V(s) \, ds$$

$$= \int_0^{t_n} (b_1 1_{(0,t_1)} + \cdots + b_n 1_{(0,t_n)})a(s)V(s) \, ds$$

with $t_1 \le \cdots \le t_n$, has a normal distribution. But this follows immediately since

$$(b_1 1_{(0,t_1)} + \cdots + b_n 1_{(0,t_n)})a(s)$$

is a piecewise continuous function.

5.6 THE WIENER PROCESS

Even with as general a definition as we have given for Gaussian processes, it is difficult to write one down based on the definition alone. It is easy to write down multivariate normal distributions, but this is not enough. The problem is to come up with distributions satisfying Kolmogorov's consistency requirements. Typically, therefore, the examination of a Gaussian process does not begin with stating its distribution function, but rather begins by assuming some conditions that force a normal joint distribution. One then adjusts the parameters so that Kolmogorov's requirements are satisfied.

We can verify by the methods of the preceding section that one such assumption is the assumption of normal independent increments. By this we mean that $X(0)$ has some fixed distribution and that the increments $X(t_{i+1}) - X(t_i)$ are normal and independent. If we make the further assumptions that

  (i) $X(0) = 0$
  (ii) $E[X(t_{i+1}) - X(t_i)] = 0$
  (iii) the variance of the increments is given by a function of the difference in the times

$$\text{var}\,[X(t_{i+1}) - X(t_i)] = c(t_{i+1} - t_i)$$

then the joint distribution is given by

$$W(x_1, t_1; \ldots; x_n, t_n) = \frac{1}{\sqrt{2\pi c(t_1)}} \exp\left[-\frac{1}{2} \frac{x_1^2}{c(t_1)} \right] \prod_{k=1}^{n-1} \frac{1}{\sqrt{2\pi c(t_{k+1} - t_k)}}$$

$$\times \exp\left[-\frac{1}{2} \frac{(x_{k+1} - x_k)^2}{c(t_{k+1} - t_k)} \right] \tag{5.6.1}$$

with $t_1 \le \cdots \le t_n$. To satisfy Kolmogorov's consistency requirements, we must have

$$\int_{-\infty}^{\infty} W(x_1, t_1; \ldots; x_n, t_n)\, dx_i = W(x_1, t_1; \ldots; x_{i-1}, t_{i-1}; x_{i+1}, t_{i+1}; \ldots; x_n, t_n)$$

which in this case holds if the variance of the sum is the sum of the variances

$$c(t_{i+1} - t_i) + c(t_i - t_{i-1}) = c(t_{i+1} - t_{i-1}) \tag{5.6.2}$$

The only positive solution to this is for c to be the linear function

$$c(t) = tB \qquad (5.6.3)$$

where B is a positive constant. Substituting this formula for c in (5.6.1) gives the joint density for what is called the Wiener process. We shall see in Chapter 7 that the Wiener process provides one description of Brownian motion, and in fact this process is sometimes referred to as the Brownian motion process.

It is interesting to find the covariance matrix for this process. Since for $t_i < t_j$,

$$
\begin{aligned}
(t_j - t_i)B &= E[(X(t_j) - X(t_i))^2] \\
&= E[(X(t_j))^2] - 2E[X(t_j)X(t_i)] + E[(X(t_i))^2] \\
&= t_j B - 2E[X(t_j)X(t_i)] + t_i B
\end{aligned}
$$

we have

$$E[X(t_j)X(t_i)] = t_i B$$

so

$$r_{ij} = B \min(t_i, t_j) \qquad (5.6.4)$$

The Wiener process as a limit of the Poisson process

Consider a Poisson process $N(t)$ with density ν. The quantity

$$X(t) = N(t) - \nu t$$

has independent increments, with mean zero and variance

$$E[(X(t) - X(s))^2] = \nu|t - s|$$

According to the central limit theorem, as $t \to \infty$, $X(t)$ becomes normally distributed. Further, defining

$$Y(t) = \frac{1}{\sqrt{\nu}}[N(t) - \nu t]$$

by a change of scale, we can conclude as $\nu \to \infty$ that the increments of $Y(t)$ become independent and normally distributed, so $Y(t)$ becomes a Wiener process, with

$$E[Y(t)] = 0$$

and

$$E[(Y(t))^2] = t$$

In Chapter 8 we shall see that the Wiener process can also be approximated in terms of random walk.

From the discussion in Chapter 4, the random noise process can be considered as the output of a linear system, where the input is the limit of the derivative of $Y(t)$. This limit is referred to as the white noise or purely random process. The white noise process does not exist as a Gaussian process in the ordinary sense, since $E[(Y(t))^2] \to \infty$ as $v \to \infty$, but can be interpreted as a stationary Gaussian process whose covariance function is singular, that is, $r(t) = \delta(t)$.

5.7 STATIONARY PROCESSES, SPECTRAL DENSITY

Stationary processes

If $X(t)$ is such that $E[(X(t))^2] < \infty$, it is called a **second-order process.** The mean and covariance functions of a second order process $X(t)$ are given by

$$\mu(t) = E[X(t)]$$

and

$$r(t, s) = E[(X(t) - \mu(t))(X(s) - \mu(t))] \qquad (5.7.1)$$

The covariance function has the obvious symmetry property

$$r(s, t) = r(t, s) \qquad (5.7.2)$$

It is also a positive definite function of two variables, meaning that for all complex numbers $z_1, \ldots, z_n$ and times $t_1, \ldots, t_n$

$$\sum_{i=1}^{n} \sum_{j=1}^{n} r(t_i, t_j) z_i \bar{z}_j \geq 0 \qquad (5.7.3)$$

which follows since this is $E\left[\left|\sum_{i=1}^{n} (X(t_i) - \mu(t_i)) z_i\right|^2\right]$. If $X(t)$ is a Gaussian process, we have seen that for a given set of times $t_1, \ldots, t_n$, the matrix $\mathbf{R} = r(t_i, t_j)$ found from the covariance function is the inverse of the matrix $\mathbf{A}$ in the joint density of $X(t_1), \ldots, X(t_n)$.

A stochastic process $X(t)$ is called **stationary** if its joint distributions are unchanged under time translations, which means that $X(t)$ and $Y(t) = X(t + \tau)$ have the same joint distributions. A Gaussian process will be stationary if its mean function is constant and its covariance function is unchanged by time translations

$$r(t + \tau, s + \tau) = r(t, s)$$

If the mean function of a second-order process is constant and the covariance function is invariant under time translations, $X(t)$ is called a **weakly stationary** stochastic process. A stationary process is weakly stationary, but there exist weakly stationary processes that are not stationary. For both stationary and weakly stationary processes the covariance function is a function of the difference of the two times

$$r(t) = r(s, s + t) \tag{5.7.4}$$

independent of the value of s. It is clear that r then has the following properties:

$$\text{(i) } r(t) = r(-t)$$
$$\text{(ii) } r(0) = \text{Var } X(t) \tag{5.7.5}$$
$$\text{(iii) } |r(t)| \leq r(0)$$

Properties (i) and (ii) are trivial, and property (iii) follows from the Cauchy–Schwarz inequality.

Recall the end of Chapter 4, where we found the spectral density of the shot noise process. For a general weakly stationary process, provided that

$$\int_{-\infty}^{\infty} r(t) \, dt < \infty$$

the spectral density is defined as the inverse Fourier transform of the covariance function

$$f(\omega) = \frac{1}{2\pi} \int_{-\infty}^{\infty} e^{-i\omega t} r(t) \, dt \tag{5.7.6}$$

A consequence of the $r(t)$ being even is

$$f(\omega) = \frac{1}{\pi} \int_{0}^{\infty} r(t) \cos \omega t \, dt \tag{5.7.7}$$

so that f is real and symmetric. In fact, by using the positive definiteness of the covariance function and approximating this integral, it is possible to show that f is nonnegative. Applying the Fourier inversion formula to (5.7.6) gives

$$r(t) = \int_{-\infty}^{\infty} e^{i\omega t} f(\omega) \, d\omega \tag{5.7.8}$$

so that if $r(0) = 1$, f is a probability density.

Characterizing covariance functions: an application of Bochner's theorem

If $X(t)$ is a weakly stationary process, the positive definiteness of the covariance function takes the form

$$\sum_{j=1}^{n} \sum_{k=1}^{n} r(t_j - t_k) z_j \bar{z}_k = E\left[\left|\sum_{j=1}^{n} (X(t_j) - \mu) z_j\right|^2\right] \geq 0$$

for any finite set of times t_j and complex numbers z_j. If we assume that $X(t)$ is mean square continuous, which means that for all t_0

$$\lim_{t \to t_0} E[(X(t) - X(t_0))^2] = 0$$

then $r(t)$ is also continuous. For any such $r(t)$ we can define a stationary mean square continuous Gaussian stochastic process which has $r(t)$ as its covariance function, so the class of continuous positive definite functions is exactly the class of covariance functions of mean square continuous weakly stationary processes. **Bochner's theorem** says that any $r(t)$ can be represented as a Steiltjes integral of the form

$$r(t) = \int_{-\infty}^{\infty} e^{i\omega t} \, dF(\omega)$$

where F is real nondecreasing and bounded

$$0 \leq F(\omega) \leq M$$

The function F is referred to as the spectral distribution function of X, despite the fact that in general $M \neq 1$. If F is differentiable, its derivative is referred to as the spectral density function of X, as above.

Approximating Gaussian processes

A stationary mean zero Gaussian process $W(t)$ can be interpreted approximately as a sum of the form

$$W(t) = X + \sum_{k=1}^{\infty} X_k \cos \omega_k t + Y_k \sin \omega_k t, \qquad \omega_k = \frac{2\pi k}{T}$$

where X, X_k and Y_k are independent mean zero normal random variables. This is sometimes referred to as Rice's representation of the process. Thus the random noise process of Chapter 4, instead of being a special type of stationary Gaussian process, can in fact approximate an arbitrary stationary Gaussian process. To see this, write $W(t)$ as a Fourier series

$$W(t) = \sum_{k=-\infty}^{\infty} \alpha(\omega_k) e^{i\omega_k t}$$

where

$$\alpha(\omega_k) = \frac{1}{T} \int_{-T/2}^{T/2} W(t) e^{-i\omega_k t}\, dt, \qquad \omega_k = \frac{2\pi k}{T}$$

Rewriting this, we obtain

$$W(t) = \left[\frac{1}{T} \int_{-T/2}^{T/2} W(s)\, ds\right] + 2 \sum_{k=1}^{\infty} \left[\frac{1}{T} \int_{-T/2}^{T/2} W(s) \cos \omega_k s\, ds\right] \cos \omega_k t$$
$$+ \left[\frac{1}{T} \int_{-T/2}^{T/2} W(s) \sin \omega_k s\, ds\right] \sin \omega_k t$$

The quantities in brackets are mean zero normal random variables, being integrals of mean zero normal random variables. Since $W(t)$ is stationary, each pair in the summation has the same distribution. Independence follows from showing the corresponding covariances are zero, which follows from the fact that $r(t)$ is an even function. For example,

$$E\left[\int_{-T/2}^{T/2} W(s) \cos \omega_k s\, ds \int_{-T/2}^{T/2} W(t) \sin \omega_k t\, dt\right]$$
$$= \int_{-T/2}^{T/2} \int_{-T/2}^{T/2} r(t-s) \cos \omega_k s \sin \omega_k t\, ds\, dt = 0$$

Thus the representation applies for $-T/2 < t < T/2$.

It applies for all t if $W(t)$ is periodic with period T. The covariance is given by

$$r(t) = E[W(t+s)W(s)]$$

$$= E[X^2] + \sum_{k=1}^{\infty} E[X_k^2] \cos \omega_k(t+s) \cos \omega_k t + E[Y_k^2] \sin \omega_k(t+s) \sin \omega_k t$$

$$= E[X^2] + \sum_{k=1}^{\infty} b_k \cos \omega_k t$$

where $b_k = E[X_k^2] = E[Y_k^2]$. Thus the spectral distribution gives an indication of the variance of the contribution that any particular frequency ω_k makes toward the process. To exactly represent a nonperiodic process we need to make the sum above an integral.

Stationary processes and linear systems

In a linear system L, it is usually easier to predict the effect that the system has on the spectral density and then later find the covariance function. This is based on the **frequency response function** $A(\omega)$ of the system, defined by

$$A(\omega) = \int_{-\infty}^{\infty} e^{i\omega s} f(s) \, ds$$

The reason it is called the frequency response functions is that

$$L(e^{-i\omega t}) = \int_{-\infty}^{\infty} e^{-i\omega(t-s)} f(s) \, ds$$

$$= A(\omega) e^{-i\omega t}$$

so $|A(\omega)|$ is the amplitude of the output when the input is $x(t) = e^{-i\omega t}$.

Using this, we can compute the spectral density $f_Y(\omega)$ of the output, if the input $X(t)$ is a weakly stationary stochastic process with covariance function $r_X(t)$. Since

$$f_Y(\omega) = \frac{1}{2\pi} \int_{-\infty}^{\infty} e^{-iwt} r_Y(t) \, dt$$

by applying the rule from Section 4.1 for calculating the moments of the output to a linear system we get

$$= \frac{1}{2\pi} \int_{-\infty}^{\infty} e^{-i\omega t} \int_{-\infty}^{\infty} \int_{-\infty}^{\infty} r_X(t + s_2 - s_1) f(s_1) f(s_2) \, ds_1 \, ds_2 \, dt$$

and interchanging the order of integration

$$= \int_{-\infty}^{\infty} e^{i\omega s_1} f(s_1) \, ds_1 \int_{-\infty}^{\infty} e^{-i\omega s_2} f(s_2) \, ds_2 \, f_X(\omega)$$

$$= A(\omega)\overline{A(\omega)} f_X(\omega)$$

This says the spectral densities are related by the simple formula

$$f_Y(\omega) = |A(\omega)|^2 f_X(\omega)$$

Hilbert space representations

Suppose that $X(t)$ is a mean zero Gaussian process. The vector space of all possible linear combinations

$$\sum_{k=1}^{n} a_k X(t_k)$$

with the inner product

$$<X, Y> = E[XY]$$

becomes a Hilbert space H, provided we add to it all the possible limits of Cauchy sequences. This Hilbert space is called the closed linear manifold determined by $X(t)$. For Gaussian processes, H is useful primarily because every element of H has a normal distribution, and its statistical properties are determined by its mean and covariances. These in turn can be described in terms of the inner product given above.

We can make the following observations. Since the random variables in H are normal and of mean zero, orthogonality,

$$E[XY] = 0$$

implies that X and Y are independent. Further, conditional expectation is equivalent to orthogonal

$$E[Y|X_1, \ldots, X_n]$$

projection onto the subspace of all linear combinations of $X_1, \ldots, X_n$. Thus this is the natural framework in which to speak about conditional expectation of Gaussian processes, as we shall see in Chapter 6.

The situation is considerably more general than this, however. The Hilbert space setting also allows us to give a brief description of the **spectral representation theorem** for weakly stationary processes. If $X(t)$ is any mean zero weakly stationary process, we can also consider the (complex) Hilbert space H which is the closed linear manifold generated by complex linear combinations of the $X(t_k)$. Suppose F is the spectral distribution function of $X(t)$. From the correspondence between H and the complex Hilbert space of functions with inner product

$$<f(\omega), g(\omega)> = \int f(\omega)\overline{g(\omega)} \, dF(\omega)$$

given by

$$X(t) \sim e^{it\omega}$$

we can draw some striking conclusions. By Bochner's theorem,

$$E[X(t)\overline{X(s)}] = \int_{-\infty}^{\infty} e^{it\omega}\overline{e^{is\omega}} \, dF(\omega)$$

so that the correspondence preserves (complex) inner products. This extends to linear combinations, meaning

$$E[X\overline{Y}] = \int_{-\infty}^{\infty} f(\omega)\overline{g(\omega)} \, dF(\omega)$$

where

$$X = \sum_{k=1}^{n} a_k X(t_k), \qquad Y = \sum_{k=1}^{n} b_k X(t_k)$$

and

$$f(\omega) = \sum_{k=1}^{n} a_k e^{it_k \omega}, \qquad g(\omega) = \sum_{k=1}^{n} b_k e^{it_k \omega}$$

In fact it can be extended to a correspondence between any element of H and complex valued functions which are square integrable with respect to F. Since it preserves inner products and hence the norm, it is called an **isometry** between the Hilbert spaces. Therefore, corresponding to the function $1_{(-\infty, a]}(\omega)$ is a random variable Z_a in H,

$$Z_a \sim 1_{(-\infty, a]}(\omega)$$

It is possible to define an integral with respect to this, which from the correspondence

$$\frac{dZ_a}{da} \sim \delta(a - \omega)$$

satisfies

$$\int_{-\infty}^{\infty} e^{ita} \, dZ_a \sim \int_{-\infty}^{\infty} e^{ita} \, \delta(a - \omega) \, da = e^{it\omega}$$

This says that

$$\int_{-\infty}^{\infty} e^{ita} \, dZ_a = X_t$$

which is the spectral representation of a stationary process. We shall not give the definition of this integral here. One can consider this to be a complete generalization of Rice's representation, since we do not require $X(t)$ to be Gaussian or periodic. This key result makes it possible to determine many of the properties of weakly stationary processes.

PROBLEMS

5.1. **(a)** Explain the reason behind assumption (ii) in Kolmogorov's consistency requirements.

   (b) Show that this condition is satisfied by the density function of a Gaussian process.

5.2. Explain why every change of coordinates that preserves the inner product, so it preserves the geometry of Euclidean space, is given by an orthogonal transformation, and vice versa.

5.3. Show that every positive definite matrix has a square root: $\mathbf{A} = \mathbf{S}^T\mathbf{S}$. Using this, and by the change of variables $\mathbf{y} = \mathbf{S}\mathbf{x}$, verify that if $\mathbf{X}$ has a multivariate normal density

$$f_{\mathbf{X}}(x_1, \ldots, x_n) = \frac{\sqrt{\det \mathbf{A}}}{\sqrt{(2\pi)^n}} \exp\left(-\tfrac{1}{2}\mathbf{x}^T\mathbf{A}\mathbf{x}\right)$$

we have

$$E[\exp(i\boldsymbol{\xi}^T\mathbf{X})] = \exp\left(-\tfrac{1}{2}\boldsymbol{\xi}^T\mathbf{A}^{-1}\boldsymbol{\xi}\right)$$

5.4. Suppose that $X(t)$ and $Y(t)$ are Gaussian processes.
(a) By using the characteristic function, verify that

$$Z(t) = X(t) + Y(t)$$

is a Gaussian process.
(b) Express the covariance function for $Z(t)$ in terms of the covariance functions

$$r_X(t, s) = E[(X(t) - \mu_X(t))(X(s) - \mu_X(s))]$$
$$r_Y(t, s) = E[(Y(t) - \mu_Y(t))(Y(s) - \mu_Y(s))]$$

and the cross covariance

$$r_{XY}(t, s) = E[(X(t) - \mu_X(t))(Y(t) - \mu_Y(t))]$$

5.5. If $X(t)$ is a mean zero Gaussian process, with continuous covariance function, and $a(t)$ is a piecewise continuous function, find the mean and covariance function of the Gaussian process

$$Z(t) = \int_0^t a(s)X(s)\, ds$$

5.6. **(a)** If X is a mean zero normal random variable, find the general form of $E[X^n]$ by using the higher correlation rule.
(b) Verify that part (a) is true, by using the form of the moment generating function of a normal random variable and

$$M_X(t) = \sum_{n=0}^{\infty} E[X^n]\frac{t^n}{n!}$$

5.7. Prove that the only solution to (5.6.2) is the linear function stated in (5.6.3).

5.8. **(a)** Find the derivative

$$\frac{\partial^2 f(t, s)}{\partial s\, \partial t}$$

of $f(t, s) = \min (t, s)$. This involves the Dirac delta function.

(b) Suppose that X is normal of mean zero and variance 1. Find the density for the 2-dimensional random variable $\mathbf{X} = (X, X)$.

5.9. In view of the fact that any symmetric real matrix can be diagonalized, it is possible to have a density of the form

$$f_{\mathbf{X}}(x_1, \ldots, x_n) = Ke^{-(1/2)\mathbf{x}^{\mathrm{T}}\mathbf{A}\mathbf{x}}$$

where $\mathbf{A}$ is symmetric with nonnegative eigenvalues and has determinant 0?

5.10. Show that if $X(t)$ is the Wiener process

$$E\left[\left(\frac{X(t) - X(s)}{t - s}\right)^2\right] \to \infty \qquad \text{as } t \to s$$

and conclude that the average kinetic energy of a particle governed by the Wiener process is infinite.

5.11. Determine whether the notion of a vector with independent components can be made independent of the choice of coordinate system by considering the following. Suppose that $\mathbf{X} = (X_1, \ldots, X_n)$ is a random vector, with independent identical symmetrically distributed components. Show that for any orthogonal transformation $\mathbf{Q}$, $\mathbf{QX}$ is also a vector with independent components if and only if $X_1, \ldots, X_n$ are independent normal random variables.

5.12. Suppose that $X(t)$ is a Gaussian process, so it satisfies the higher correlation rule for any t. Show that

$$\int_0^t a(s)X(s)\, ds$$

satisfies the higher correlation rule and thus provide an alternative proof that this integral is a normal random variable.

5.13. Suppose that $X(t)$ is stationary with covariance function $r(t)$. Show that

(a) $\text{var } (X(t + s) - X(s)) = 2(r(0) - r(t))$

(b) $P[|X(t + s) - X(s)| \geq k] \leq \dfrac{2}{k^2} (r(0) - r(t))$

5.14. If $W(t)$ is the Wiener process, show that

$$X(t) = \frac{W(t + \epsilon) - W(t)}{\epsilon}$$

is a stationary Gaussian process, by showing that its covariance function is unchanged by time translations. What is the limit of the covariance as $\epsilon \to 0$? What can we conclude from this about the derivative of the Wiener process?

5.15. Show that if $X(t)$ is Gaussian, and f and g are real-valued functions, then $f(t)X(g(t))$ is Gaussian. Find its mean and covariance functions.

5.16. Suppose that X and Y are jointly normal random variables. Use the higher correlation rule to show that

$$E[X^2Y^2] = E[X^2][Y^2] + 2E[XY]$$

Use this to find the covariance function for $W^2(t)$, where $W(t)$ is the Wiener process.

5.17. If $W(t)$ is the Wiener process, show that the following are all Wiener processes
  (a) $aW(t/a^2)$, $a \neq 0$.
  (b) $W(t + a) - W(a)$.
  (c) $tW(1/t)$, where it is taken to be 0 at $t = 0$.

5.18. Suppose that $X(t)$ is a stationary Gaussian process. Show that

$$\int_t^{t+b} X(s)\, ds = Y(t)$$

is a stationary Gaussian process.

5.19. Find the spectral density function for $X(t) = N(t + s) - N(t)$, where $N(t)$ is a Poisson process of parameter v.

5.20. A stochastic process $X(t)$ is called continuous in mean square if

$$\lim_{t \to a} E[(X(t) - X(a))^2] = 0$$

Show that $X(t)$ is continuous in mean square if its mean and covariance functions are continuous functions.

5.21. Suppose that $X(t)$ and $Y(t)$ are mean zero stationary processes. Let $Z(t) = X(t) + Y(t)$. Under what conditions is the spectral density of $Z(t)$ the sum of the spectral densities of $X(t)$ and $Y(t)$?

5.22. What is the frequency response function for the linear system

$$L(X(t)) = \frac{dX(t)}{dt} ?$$

5.23. Suppose that Z_n, $n = 1, 2, \ldots$ is a sequence of normal random variables, and that Z_n converges to a random variable Z in mean square, meaning that

$$\lim_{n \to \infty} E[(Z - Z_n)^2] = 0$$

(a) Show that Z is a normal random variable.

(b) Conclude that the Hilbert space formed by all linear combinations of a family of normal random variables and their mean square limits contains only normal random variables.

5.24. Suppose that $X(t)$ is a mean zero Gaussian process. Consider the vector space of finite linear combinations

$$\mathbb{V} = \left\{ \sum a_i X(t_i) \right\}$$

(a) Verify that $\langle V_1, V_2 \rangle = \text{cov}(V_1, V_2)$ defines a real inner product on this vector space by showing for any $V_1, V_2, V_3 \in \mathbb{V}$, and real a, b:

(i) $\langle aV_1, bV_2 \rangle = ab\langle V_1, V_2 \rangle$

(ii) $\langle V_1, V_2 + V_3 \rangle = \langle V_1, V_2 \rangle + \langle V_1, V_3 \rangle$

(iii) $\langle V_1, V_2 \rangle = \langle V_2, V_1 \rangle$

(b) With this inner product, show that the random variable

$$E[X(t_n)|X(t_1), \ldots, X(t_n)]$$

is the projection of $X(t_n)$ onto the span of $X(t_1), \ldots, X(t_n)$.

CHAPTER

6

Markov Gaussian Processes

One of the best known of all types of stochastic processes with a continuum of states are those Gaussian processes which are stationary and Markovian. These are called the Ornstein–Uhlenbeck processes. In this chapter we show Doob's theorem, which states that a stationary Gaussian process $X(t)$ is an Ornstein-Uhlenbeck process if and only if its covariance has an exponential form. More generally, if $X(t)$ is not assumed stationary, the Markov property is shown to be equivalent to a simple formula involving the covariance function of the process.

6.1 MARKOV PROCESSES

A stochastic process $X(t)$ is called a Markov process if for any n and $t_1 < t_2 \ldots < t_{n+1}$ we have

$$P[X(t_{n+1}) \leq a | X(t_n) \leq x_n, \ldots, X(t_1) \leq x_1] = P[X(t_{n+1}) \leq a | X(t_n) \leq x_n]$$

(6.1.1)

which is often imprecisely expressed in words as "the future is independent of the past given the present." Suppose that $X(t)$ is a real-valued process with continuous joint density W and $t_1 < \cdots < t_n < t_{n+1} < \cdots < t_m$. The conditional density of $X(t_m), \ldots, X(t_{n+1})$ given $X(t_n) = x_n, \ldots, X(t_1) = x_1$ is defined in terms of the joint density $W(x_1, t_1; \ldots; x_n, t_n)$ by

$$P(x_m, t_m; \ldots; x_{n+1}, t_{n+1} | x_n, t_n; \ldots; x_1, t_1) = \frac{W(x_1, t_1; \ldots; x_m, t_m)}{W(x_1, t_1; \ldots; x_n, t_n)}$$

(6.1.2)

and in this case the process will be Markovian if

$$P(x_{n+1}, t_{n+1} | x_n, t_n; \ldots; x_1, t_1) = P(x_{n+1}, t_{n+1} | x_n, t_n) \qquad (6.1.3)$$

This is equivalent to

$$P(x_{n+1}, t_{n+1}; \ldots; x_2, t_2 | x_1, t_1) = P(x_{n+1}, t_{n+1} | x_n, t_n) \ldots P(x_2, t_2 | x_1, t_1)$$
$$(6.1.4)$$

which is sometimes taken as an alternative definition of the Markov property. The equivalence of these two equations is easily demonstrated. By the definition of conditional density, (6.1.3) becomes

$$\frac{W(x_1, t_1; \ldots; x_{n+1}, t_{n+1})}{W(x_1, t_1; \ldots; x_n, t_n)} = \frac{W(x_n, t_n; x_{n+1}, t_{n+1})}{W(x_n, t_n)}$$

or

$$\frac{W(x_1, t_1; \ldots; x_{n+1}, t_{n+1})}{W(x_1, t_1)} = \frac{W(x_n, t_n; x_{n+1}, t_{n+1})}{W(x_n, t_n)} \frac{W(x_1, t_1; \ldots; x_n, t_n)}{W(x_1, t_1)}$$

Rewriting this in terms of conditional densities

$$P(x_{n+1}, t_{n+1}; \ldots; x_2, t_2 | x_1, t_1) = P(x_{n+1}, t_{n+1} | x_n, t_n) P(x_n, t_n; \ldots; x_2, t_2 | x_1, t_1)$$

from which (6.1.4) follows by induction.

6.2 DOOB'S THEOREM

If $X(t)$ is a stationary Gaussian process, with mean zero and variance one, then it will be Markovian if and only if its covariance function satisfies

$$r(t) = e^{-\gamma|t|}, \qquad \gamma > 0 \qquad (6.2.1)$$

This result is known as Doob's theorem. We prove here this is a necessary condition, and sufficiency is shown in Section 6.4.

By (6.1.4), the Markov property implies that

$$W(x_1, t_1; x_2, t_2; x_3, t_3) W(x_2, t_2) = W(x_1, t_1; x_2, t_2) W(x_2, t_2; x_3, t_3) \qquad (6.2.2)$$

and these four densities can be easily computed. Since $X(t)$ has mean zero and variance 1,

$$W(x_2, t_2) = \frac{1}{\sqrt{2\pi}} e^{-(1/2)x_2^2} \qquad (6.2.3)$$

The two-dimensional density $W(x_1, t_1; x_2, x_2)$ is

$$W(x_1, t_1; x_2, t_2) = \frac{\sqrt{\det A}}{2\pi} e^{-(1/2)(a_{11}x_1^2 + 2a_{12}x_1x_2 + a_{22}x_2^2)}$$

where A is the inverse of the covariance matrix, which is given by the covariance function

$$
\begin{aligned}
A = (a_{ij}) &= (r(t_i - t_j))^{-1} \\
&= \begin{bmatrix} r(0) & r(t_1 - t_2) \\ r(t_2 - t_1) & r(0) \end{bmatrix}^{-1}
\end{aligned}
$$

or setting $r_{ij} = r(t_i - t_j)$, recalling $r(t)$ is even and $r(0) = \mathrm{var}\, X(t)$,

$$
= \begin{bmatrix} 1 & r_{12} \\ r_{12} & 1 \end{bmatrix}^{-1}
$$

$$
= \begin{bmatrix} \dfrac{1}{1 - r_{12}^2} & \dfrac{-r_{12}}{1 - r_{12}^2} \\[2ex] \dfrac{-r_{12}}{1 - r_{12}^2} & \dfrac{1}{1 - r_{12}^2} \end{bmatrix}
$$

Therefore

$$W(x_1, t_1; x_2, t_2) = \frac{1}{2\pi\sqrt{1 - r_{12}^2}} \exp\left(-\frac{1}{2} \frac{x_1^2 - 2r_{12}x_1x_2 + x_2^2}{1 - r_{12}^2} \right) \tag{6.2.4}$$

and

$$W(x_2, t_2; x_3, t_3) = \frac{1}{2\pi\sqrt{1 - r_{23}^2}} \exp\left(-\frac{1}{2} \frac{x_2^2 - 2r_{23}x_2x_3 + x_3^2}{1 - r_{23}^2} \right)$$

Also,

$$W(x_1, t_1; x_2, t_2; x_3, t_3) = \frac{\sqrt{\det A}}{\sqrt{(2\pi)^3}} \exp\left(-\tfrac{1}{2} x^T A x \right)$$

where

$$A = \begin{bmatrix} 1 & r_{12} & r_{13} \\ r_{12} & 1 & r_{23} \\ r_{13} & r_{23} & 1 \end{bmatrix}^{-1}$$

so that

$$\det \mathbf{A} = \frac{1}{1 - r_{23}^2 - r_{12}^2 - r_{13}^2 + 2r_{12}r_{13}r_{23}}$$

Substituting these densities into (6.2.2) with $x_1 = x_2 = x_3 = 0$, we get

$$(1 - r_{12}^2)(1 - r_{23}^2) = 1 - r_{23}^2 - r_{12}^2 - r_{31}^2 + 2r_{12}r_{13}r_{23}$$

or

$$(r_{13} - r_{12}r_{23})^2 = 0$$

or

$$r_{12}r_{23} = r_{13} \tag{6.2.5}$$

This, by setting $s = t_2 - t_1$, $t = t_3 - t_2$, becomes

$$r(s)r(t) = r(s + t) \tag{6.2.6}$$

To show that this functional equation is equivalent to (6.2.1), set $s = t$ to get

$$r^2(t) = r(2t)$$

and if $s = (n - 1)t$

$$r((n - 1)t)r(t) = r(nt)$$

so that induction gives

$$r^n(t) = r(nt) \tag{6.2.7}$$

Setting $t = 1/k$ and $n = k$ gives

$$r(1/k) = r^{1/k}(1), \qquad k = 1, 2, \ldots$$

which with (6.2.7) yields

$$r(n/k) = r^n(1/k) = r^{n/k}(1), \qquad k, n = 1, 2, \ldots$$

Noticing that $r(1) < 1$,

$$r(n/k) = e^{-\gamma n/k}, \qquad \gamma > 0$$

and assuming $r(t)$ is continuous, since $r(0) > 0$, we can conclude

$$r(t) = e^{-\gamma|t|}, \qquad \gamma > 0$$

6.3   THE CHAPMAN–KOLMOGOROV EQUATION

The joint density of a Markov process, by (6.1.4) satisfies

$$W(x_1, t_1; \ldots; x_{n+1}, t_{n+1}) = P(x_{n+1}, t_{n+1}|x_n, t_n) \cdots P(x_2, t_2|x_1, t_1)W(x_1, t_1) \tag{6.3.1}$$

for $t_1 < \cdots < t_{n+1}$, and in this case the consistency condition (5.0.3) is equivalent to

$$\int_{-\infty}^{\infty} P(x_{i+1}, t_{i+1}|x_i, t_i)P(x_i, t_i|x_{i-1}, t_{i-1}) \, dx_i = P(x_{i+1}, t_{i+1}|x_{i-1}, t_{i-1}) \tag{6.3.2}$$

which is often referred to as the Chapman–Kolmogorov or Einstein–Smoluchowski equation. As an application of this relation, we now give an alternative proof of the necessity of (6.2.1) in Doob's theorem.

By (6.2.3) and (6.2.4) we know that

$$P(x_3, t_3|x_1, t_1) = \frac{1}{\sqrt{2\pi(1 - r_{13}^2)}} \exp\left[-\frac{1}{2} \frac{(x_3 - r_{13}x_1)^2}{1 - r_{13}^2} \right] \tag{6.3.3}$$

which as a function of x_3 is a normal density with mean $r_{13}x_1$, so

$$r_{13}x_1 = \int_{-\infty}^{\infty} x_3 P(x_3, t_3|x_1, t_1) \, dx_3 \tag{6.3.4}$$

By the Chapman–Kolmogorov equation this is

$$r_{13}x_1 = \int_{-\infty}^{\infty} x_3 \left[\int_{-\infty}^{\infty} P(x_3, t_3|x_2, t_2)P(x_2, t_2|x_1, t_1) \, dx_2 \right] dx_3$$

$$= \int_{-\infty}^{\infty} r_{23}x_2 P(x_2, t_2|x_1, t_1) \, dx_2$$

$$= r_{23}r_{12}x_1$$

or, for $x_1 \neq 0$,

$$r_{13} = r_{12}r_{23}$$

This is the familiar relation of Doob's theorem. The Chapman–Kolmogorov equation is used again in Section 7.4 to derive the Fokker–Planck equation.

6.4 COMPLETION OF DOOB'S THEOREM

Any family of probability densities satisfying (6.3.1) and (6.3.2) defines a Markov process. Therefore, we can complete the proof of Doob's theorem by verifying these two equations for the family of mean zero Gaussian densities whose covariance matrix **R** is given by

$$(r_{ij}) = (e^{-\gamma|t_i - t_j|}) \tag{6.4.1}$$

Actually, this will show a little more than is required, since it demonstrates the existence of stationary Gaussian Markov processes, which so far has been assumed.

To avoid notational complexity we show (6.3.1) for three times only, but the proof in the general case is similar. In this case it takes the form

$$W(x_1, t_1; x_2, t_2; x_3, t_3) = P(x_3, t_3|x_2, t_2)P(x_2, t_2|x_1, t_1)W(x_1, t_1)$$

which is

$$\frac{\sqrt{\det \mathbf{A}}}{\sqrt{(2\pi)^3}} e^{-(1/2)\mathbf{x}^T\mathbf{A}\mathbf{x}} = \frac{1}{\sqrt{(2\pi)^3(1 - r_{12}^2)(1 - r_{23}^2)}}$$
$$\times \exp\left[-\frac{1}{2}\left(x_1^2 + \frac{(x_3 - r_{23}x_2)^2}{1 - r_{23}^2} + \frac{(x_2 - r_{12}x_1)^2}{1 - r_{12}^2}\right)\right] \tag{6.4.2}$$

where $\mathbf{A} = \mathbf{R}^{-1}$ and **R** is given by (6.4.1). The exponent of the right-hand side can be written $-(1/2)\mathbf{x}^T\mathbf{B}\mathbf{x}$, where

$$b_{11} = 1 + \frac{r_{12}^2}{1 - r_{12}^2}$$

$$b_{12} = b_{21} = \frac{-r_{12}}{1 - r_{12}^2}$$

$$b_{13} = b_{31} = 0$$

$$b_{22} = \frac{1}{1 - r_{12}^2} + \frac{r_{23}^2}{1 - r_{23}^2}$$

$$b_{23} = b_{32} = \frac{-r_{23}}{1 - r_{23}^2}$$

$$b_{33} = \frac{1}{1 - r_{23}^2}$$

One can easily check that $\mathbf{BR} = \mathbf{I}$ by making use of the exponential relation (6.2.5). For instance, the product of the second row of $\mathbf{B}$ with the third column of $\mathbf{R}$ is

$$b_{21}r_{13} + b_{22}r_{23} + b_{23}r_{33}$$

$$= \frac{-r_{12}r_{13}}{1 - r_{12}^2} + \frac{r_{23}}{1 - r_{12}^2} + \frac{r_{23}^3}{1 - r_{23}^2} + \frac{-r_{23}r_{33}}{1 - r_{23}^2}$$

using $r_{12}r_{23} = r_{13}$ and $r_{33} = 1$

$$= \frac{-r_{12}^2 r_{23}}{1 - r_{12}^2} + \frac{r_{23}}{1 - r_{12}^2} + \frac{r_{23}^3}{1 - r_{23}^2} + \frac{-r_{23}}{1 - r_{23}^2}$$

$$= \frac{r_{23}(1 - r_{12}^2)}{1 - r_{12}^2} + \frac{r_{23}(r_{23}^2 - 1)}{1 - r_{23}^2}$$

$$= 0$$

as expected. The rest of the products involved are similar. Thus we have $\mathbf{A} = \mathbf{B}$, which implies equation (6.4.2).

To verify (6.3.2) we need to show that

$$\int_{-\infty}^{\infty} \frac{1}{\sqrt{(2\pi)^2(1 - r_{12}^2)(1 - r_{23}^2)}} \exp\left[-\frac{1}{2}\left(\frac{(x_2 - r_{12}x_1)^2}{1 - r_{12}^2} + \frac{(x_3 - r_{23}x_2)^2}{1 - r_{23}^2}\right)\right] dx_2$$

$$= \frac{1}{\sqrt{(2\pi)(1 - r_{13}^2)}} \exp\left[-\frac{1}{2}\left(\frac{(x_3 - r_{13}x_1)^2}{1 - r_{13}^2}\right)\right] \tag{6.4.3}$$

This can be done with the aid of (6.2.5), by completing the square of the exponent, and this finishes the proof of Doob's theorem.

6.5   THE ORNSTEIN–UHLENBECK PROCESS

Stationary Gaussian Markov processes are called Ornstein–Uhlenbeck processes. The joint density of a general Ornstein–Uhlenbeck process $X(t)$ can be expressed as

$$\frac{\sigma^2 \sqrt{\det \mathbf{A}}}{\sqrt{(2\pi)^2}} \exp\left[-\tfrac{1}{2}\sigma^2(\mathbf{x} - \boldsymbol{\mu})^{\mathrm{T}} \mathbf{A}(\mathbf{x} - \boldsymbol{\mu})\right] \tag{6.5.1}$$

where

$$\mathbf{A} = (\rho(t_i - t_j))^{-1}, \qquad \sigma^2 = \mathrm{var}\, X(t), \qquad \boldsymbol{\mu} = (\mu, \ldots, \mu)^{\mathrm{T}}, \qquad \mu = E[X(t)]$$

and

$$\rho(t) = e^{-\gamma|t|}$$

is the correlation coefficient of $X(t)$, so that the covariance function is

$$r(t) = \sigma^2 e^{-\gamma|t|} \tag{6.5.2}$$

Thus every statement about normalized (mean zero, variance 1) stationary Gaussian processes can be rephrased for general Ornstein–Uhlenbeck processes in terms of the mean μ, the variance σ^2, and the correlation coefficient $\rho(t)$.

Suppose that $X(t)$ is a normalized Ornstein–Uhlenbeck process. We can compare this process with the Wiener process by finding the distribution of an increment $Z(t - s) = X(t) - X(s)$, $s < t$. Since Z is a linear combination of two jointly normally distributed random variables, it has normal distribution, with mean

$$E[X(t) - X(s)] = E[X(t)] - E[X(s)] = 0$$

and variance

$$\begin{aligned}
E[(X(t) - X(s))^2] &= E[X^2(t)] - 2E[X(t)X(s)] + E[X^2(s)] \\
&= 2(1 - \rho(t - s)) \\
&= 2(1 - e^{-\gamma|t - s|})
\end{aligned}$$

We note that there is a limiting variance as $t \to \infty$. In view of (6.5.1) it is easy to find the conditional distribution of Z,

$$P[z, t - s | x, s] = \frac{1}{\sqrt{2\pi(1 - r^2)}} \exp\left[-\frac{1}{2}\frac{(z + (1 - r)x)^2}{1 - r^2}\right] \tag{6.5.3}$$

where $r = e^{-\gamma|t - s|}$. Thus it is clear that the increments are not independent, in contrast to the Wiener process of Section 5.6. In fact, we can see that this process is "center seeking" since the mean of the conditional distribution is $-(1 - r)x$, and $0 < (1 - r) < 1$.

6.6   CONDITIONAL EXPECTATION AND DOOB'S THEOREM

If $t_1 < \cdots < t_n$, then the conditional expectation of $X(t_n)$ given $X(t_{n-1}) = x_{n-1}$, $\ldots$, $X(t_1) = x_1$ is defined by

$$\begin{aligned}
E[X(t_n)|X(t_{n-1}) &= x_{n-1}; \ldots; X(t_1) = x_1] \\
&= \int_{-\infty}^{\infty} x_n P(x_n, t_n | x_{n-1}, t_{n-1}; \ldots; x_1, t_1)\, dx_n
\end{aligned} \tag{6.6.1}$$

For mean zero Gaussian processes the conditional expectation can be characterized as the unique linear function

$$E[X(t_n)|X(t_{n-1}) = x_{n-1}; \ldots ; X(t_1) = x_1] = b_{n-1}x_{n-1} + \cdots + b_1 x_1 \qquad (6.6.2)$$

such that the corresponding random variable

$$X(t_n) - b_{n-1}X(t_{n-1}) - \cdots - b_1 X(t_1)$$

is independent of $X(t_{n-1}), \ldots , X(t_1)$. We saw this result for jointly normal random variables in Chapter 0. It applies here since the random variables $X(t_k)$ are jointly normally distributed. We use this characterization to show the following result.

For a mean zero Gaussian process $X(t)$ to be Markovian, each of the following is necessary and sufficient:

(i) $E[X(t_n)|X(t_{n-1}) = x_{n-1}; \ldots ; X(t_1) = x_1] = E[X(t_n)|X(t_{n-1}) = x_{n-1}]$

(ii)   For $w < t < s$ the correlation coefficient $\rho(t, s) = r(t, s)/\sigma_t\sigma_s$ satisfies

$$\rho(w, s) = \rho(w, t)\rho(t, s)$$

We note that (ii) is a generalization of Doob's theorem, and that $X(t)$ is not necessarily stationary. To show this result, first note that for a Markov process (i) must be satisfied. If (i) is true, then substituting (6.6.2) in (i) gives $b_1 = \cdots b_{n-2} = 0$. Hence $P(x_n, t_n|x_{n-1}, t_{n-1}; \ldots ; x_1, t_1)$ is the density of a normal random variable with mean $b_{n-1}x_{n-1}$, whose variance is the variance of $X(t_n) - b_{n-1}X(t_{n-1})$. Since this is also the description of $P(x_n, t_n|x_{n-1}, t_{n-1})$, we have

$$P(x_n, t_n|x_{n-1}, t_{n-1}; \ldots ; x_1, t_1) = P(x_n, t_n|x_{n-1}, t_{n-1})$$

Thus (i) is equvalent to having $X(t)$ Markovian. We now show that (i) and (ii) are equivalent.

To show that (i) implies (ii), first compute

$$E\left[\left(X(t_n) - \frac{r(t_{n-1}, t_n)}{\text{var } X(t_{n-1})} X(t_{n-1}) \right) X(t_{n-1}) \right] = 0$$

so that $X(t_n) - (r(t_{n-1}, t_n)/\text{var } X(t_{n-1}))X(t_{n-1})$ and $X(t_{n-1})$ are independent. Therefore, by the characterization of the conditional expectation given above,

$$E[X(t_n)|X(t_{n-1}) = x_{n-1}] = \frac{r(t_{n-1}, t_n)}{\text{var } X(t_{n-1})} x_{n-1} \qquad (6.6.3)$$

If (i) holds, this characterization also implies that $X(t_n) - (r(t_{n-1}, t_n)/\text{var } X(t_{n-1}))X(t_{n-1})$ is independent of $X(t_{n-1}), \ldots, X(t_1)$, and in particular independent of $X(t_{n-2})$, so

$$E\left[\left(X(t_n) - \frac{r(t_{n-1}, t_n)}{\text{var } X(t_{n-1})} X(t_{n-1})\right)X(t_{n-2})\right] = 0$$

or

$$r(t_{n-2}, t_n) - \frac{r(t_{n-2}, t_{n-1})r(t_{n-1}, t_n)}{\text{var } X(t_{n-1})} = 0$$

This is the same as

$$\rho(t_{n-2}, t_n) = \rho(t_{n-2}, t_{n-1})\rho(t_{n-1}, t_n)$$

which yields (ii) by substituting $t_{n-2} = w$, $t_{n-1} = t$, $t_n = s$.

To show that (ii) implies (i), note that by (ii)

$$\rho(t_k, t_n) = \rho(t_k, t_{n-1})\rho(t_{n-1}, t_n), \qquad 1 \le k \le n - 1$$

or

$$E\left[\left(X(t_n) - \frac{r(t_{n-1}, t_n)}{\text{var } X(t_{n-1})} X(t_{n-1})\right)X(t_k)\right] = 0, \qquad 1 \le k \le n - 1$$

so that $X(t_n) - (r(t_{n-1}, t_n)/\text{var } X(t_{n-1}))X(t_{n-1})$ is independent of $X(t_1), \ldots, X(t_{n-1})$ and hence by the characterization of conditional expectation,

$$E[X(t_n)|X(t_{n-1}) = x_{n-1}; \ldots; X(t_1) = x_1] = \frac{r(t_{n-1}, t_n)}{\text{var } X(t_{n-1})} x_{n-1}$$

which by (6.6.3)

$$= E[X(t_n)|x_{n-1}, t_{n-1}]$$

Let us put this result in the geometric framework of Chapter 5. Consider the Hilbert space generated by all linear combinations of the random variables $X(t_k)$. Recall that the conditional expectation

$$E[X(t_n)|X(t_{n-1}); \ldots; X(t_1)]$$

is the projection onto the subspace of all linear combinations of $X(t_{n-1}), \ldots, X(t_1)$. Also notice that the correlation coefficient can be interpreted as the cosine of the angle between two elements of H, since

$$\rho(t, s) = \frac{<X(t), X(s)>}{\|X(t)\| \, \|X(s)\|}$$

where $<X(t), X(s)> = E[X(t)X(s)]$.

With this interpretation, (i) and (ii) are each statements that each increment $X(t_n) - X(t_{n-1})$ is a sum

$$X(t_n) - X(t_{n-1}) = aX(t_{n-1}) + Y$$

where Y is a random variable and a is a constant each of which can depend on t_n and t_{n-1} such that Y is orthogonal to every $X(s)$, $s < t_{n-1}$.

PROBLEMS

6.1. Complete the proof of the equivalence of the two forms of the Markov property (6.1.3) and (6.1.4).

6.2. **(a)** Show that any exponential random variable X is "memoryless" in the sense that

$$P[X > s + t | X > t] = P[X > t]$$

(b) By setting $g(x) = P[X > x]$, and showing that $g(s)g(t) = g(s + t)$, show that every memoryless random variable is exponential.

6.3. **(a)** Verify that the Poisson process is a Markov process, by showing that any process with independent increments is Markovian.

(b) Show that the Poisson process is the unique nondecreasing nonnegative integer valued Markov process with identically distributed independent increments.

6.4. Assuming that $r(t)$ is differentiable, show that $r(t)r(s) = r(s + t)$, $s, t > 0$ implies $r(t) = e^{kt}$, $k \neq 0$. Use this and the properties of the covariance function to show that $r(t) = e^{-\gamma|t|}$.

6.5. Suppose that the covariance function satisfies $r(t)r(s) = r(s + t)$

(a) Show that if $r(t)$ is continuous at 0, then it is continuous everywhere.

(b) Find a function satisfying $r(t)r(s) = r(s + t)$ which is not $r(t) = e^{-\alpha t}$.

6.6. Suppose that $X(t)$ is the Wiener process and $Y(t) = X(e^{2t})e^{-t}$.

(a) Show that $Y(t)$ is a stationary Gaussian process.

(b) Show it is Markovian.

(c) Conclude that it is an Ornstein–Uhlenbeck process.

(d) Find the inverse of this transformation.

6.7. Show that the "moving average" process given by

$$Z(t) = \int_t^{t+a} X(t) \, dt$$

where $X(t)$ is a Markov Gaussian process, is not Markovian.

6.8. Show that the integral

$$Z(t) = \int_0^t X(s) \, ds$$

where $X(t)$ is a Markov Gaussian process, is not Markovian.

6.9. Suppose that $X(t)$ and $Y(t)$ are independent mean zero stationary Markov processes. Under what conditions is the sum

$$Z(t) = X(t) + Y(t)$$

Markovian?

6.10. Verify the result given in (6.3.4) by computing the conditional expectation.

6.11. Check that the matrix product $\mathbf{BR} = \mathbf{I}$ of Section 6.4 holds.

6.12. If $X(t)$ is Gaussian with covariance (6.4.1), provide a general proof of equation (6.3.1).

6.13. Verify that the Kolmogorov consistency (6.4.3) holds.

6.14. Show that the Wiener process is Markovian by showing the relation

$$\rho(s, w) = \rho(s, t)\rho(t, w)$$

6.15. For the mean zero Ornstein–Uhlenbeck process with covariance function $r(t) = e^{-|t|}$, find the form of the density, and by integration demonstrate that the appropriate normal distribution is in fact stationary.

6.16. Find the form of the spectral density for a stationary Markovian Gaussian process.

6.17. If $X(t)$ is a Markov process, show that $f(t)X(t)$ is a Markov process, where f is a real-valued function.

6.18. What is the form of the covariance if the process in Problem 6.9 is conditioned to start at zero?

CHAPTER

7

Brownian Motion

Brownian motion is a real physical phenomenon, and it is relatively easy to observe. Despite this, almost 100 years passed after Brownian motion was first reported before it was quantitatively understood as a stochastic process.

An English botanist named Robert Brown reported in 1827 that plant pollen placed in water dispersed into a great number of tiny grains which displayed an irregular "swarming" motion. Other tiny particles displayed the same behavior, and he believed that this showed all matter is composed of "primitive molecules" the size of these particles. Later investigators found that molecules are actually much smaller, and they devised a number of other possible explanations. These included (1) irregular heating of the particles by light bombardment, (2) evaporation of the water, (3) electrical forces, and (4) some force between the particles. All of these possibilities were disproved by various experiments during the nineteenth century.

Experimenters were able to observe variations in the average speed of the particle when certain physical quantities were changed. The motion was faster with increasing temperature, decreasing particle size, or decreasing fluid viscosity, but it was not until 1877 that Delsaux proposed that the motion results from impacts of molecules of the fluid on the particle. This view did not immediately win general acceptance. Although Ramsay argued in 1892 that it could explain departures from established laws of osmotic pressure, as late as 1898 Quincke still believed in the temperature difference explanation. The most precise experimental studies were done by Gouy and Exner, but the particle motion is so irregular that Exner's velocity estimates were much too low and did not successfully confirm the molecular impact explanation.

Thus, by 1905, Brownian motion was well known to be very difficult to understand clearly. With this background it is easier to appreciate the impact of Einstein's theory, published the same year as his special theory of relativity. From the insights provided by Einstein, the theory developed rapidly to its present form, and we shall now work out the details of this theory.

7.1   EINSTEIN'S APPROACH

Einstein proposed that a coordinate of the position of the Brownian particle is a stochastic process $X(t)$, satisfying:

(i) For $t_1 < t_2 < \cdots < t_n$ the increments $X(t_{i+1}) - X(t_i)$ are an independent family.

(ii) The conditional density $P(x_{i+1}, t_{i+1}|x_i, t_i)$ depends only on the differences $x_{i+1} - x_i$ and $t_{i+1} - t_i$.

(iii)
$$\lim_{t_{i+1} \to t_i} \frac{1}{t_{i+1} - t_i} \int_{-\infty}^{\infty} (x_{i+1} - x_i)^k P(x_{i+1}, t_{i+1}|x_i, t_i) \, dx_{i+1}$$
$$= \begin{cases} B, & k = 2 \\ 0, & k = 1, 3, 4, \ldots \end{cases}$$

He showed that these conditions imply that $X(t)$ is a Markov process, and that the conditional densities $P(x_{i+1}, t_{i+1}|x_i, t_i)$ must satisfy a partial differential equation, which can be solved, yielding

$$P(x_{i+1}, t_{i+1}|x_i, t_i) = \frac{1}{\sqrt{2\pi B(t_{i+1} - t_i)}} \exp\left[-\frac{1}{2} \frac{(x_{i+1} - x_i)^2}{B(t_{i+1} - t_i)} \right]$$

so $X(t)$ is the Wiener process. We show this in Section 7.4. He was also able to show, based on physical principles, that

$$B = \frac{2RT}{N\gamma}$$

so that

$$E[(X(t) - X(0))^2] = \frac{2RTt}{N\gamma} \tag{7.1.1}$$

where R is the universal gas constant, T the absolute temperature, N Avogadro's number, and γ the coefficient of friction. Equation (7.1.1), which is now called Einstein's formula, made possible the determination of Avogadro's number from Brownian motion experiments, an achievement for which Perrin received the Nobel prize in 1926.

The Wiener process was recognized as an approximation to Brownian motion, since property (i) can only be satisfied approximately by a physical system. In fact, a particle whose displacement is given by this process can be shown to have infinite instantaneous velocity which is impossible for a physical system. Even so, early experiments verified properties of Brownian motion predicted from properties of the Wiener process, since the difference between the two fell within experimental error.

In the next sections we introduce another approach to Brownian motion, using the method of Ornstein–Uhlenbeck, which also gives (7.1.1).

7.2   THE VELOCITY OF THE BROWNIAN PARTICLE

The subject of this section goes back to 1908, when Paul Langevin in a short note in the 146th volume of *Comptes Rendus* proposed a way to treat Brownian motion of a free particle in a fluid.

The slow one-dimensional motion of a particle of mass M in a fluid is described according to Newton's second law, by

$$M \frac{dv}{dt} + \gamma v = 0 \qquad (7.2.1)$$

where v is the velocity and γ is the friction coefficient. Langevin treats (7.2.1) as an average equation and modifies it by adding a random "fluctuating" force $F(t)$ to represent the effect of molecular bombardment not already accounted for in the linear friction term. He thus replaces (7.2.1) by the equation now bearing his name,

$$M \frac{dv}{dt} + \gamma v = F(t) \qquad (7.2.2)$$

where $F(t)$ is interpreted as a stationary stochastic process having mean zero and singular covariance

$$E[F(t)F(s)] = 2D\delta(t - s) \qquad (7.2.3)$$

and D is to be determined later.

Treating (7.2.2) as if it were an ordinary differential equation with initial condition $V(0) = v_0$ we obtain, formally,

$$V(t) = v_0 \exp\left(-\frac{\gamma}{M} t\right) + \frac{1}{M} \int_0^t \exp\left[-\frac{\gamma}{M}(t - \tau)\right] F(\tau)\, d\tau \qquad (7.2.4)$$

This may be interpreted physically as the result of a frictional force, proportional to velocity, acting on a particle which has initial velocity v_0, and is subject to many small random independent impulses $\Delta(MV)$. Given $V(0) = v_0$, $V(t)$ has mean

$$E[V(t)] = v_0 \exp\left(-\frac{\gamma}{M} t\right) + \frac{1}{M} \int_0^t \exp\left[-\frac{\gamma}{M}(t - \tau)\right] E[F(\tau)]\, d\tau$$
$$= v_0 \exp\left(-\frac{\gamma}{M} t\right) \qquad (7.2.5)$$

and variance

$$
E\left[\left(V(t) - v_0 \exp\left(-\frac{\gamma}{M} t\right)\right)^2\right]
$$

$$
= E\left[\left(\frac{1}{M} \int_0^t \exp\left(-\frac{\gamma}{M}(t - \tau)\right) F(\tau)\, d\tau\right)^2\right] \tag{7.2.6}
$$

$$
= \left(\frac{1}{M}\right)^2 \int_0^t \int_0^t \exp\left(-\frac{\gamma}{M}(t - \eta)\right) \exp\left(-\frac{\gamma}{M}(t - \tau)\right) E[F(\eta)F(\tau)]\, d\eta\, d\tau
$$

by (7.2.3)

$$
= \frac{2D}{M^2} \exp\left(-2\frac{\gamma}{M} t\right) \int_0^t \int_0^t \exp\left(\frac{\gamma}{M}\eta\right) \exp\left(\frac{\gamma}{M}\tau\right) \delta(\tau - \eta)\, d\eta\, d\tau
$$

and using the property of δ in (1.4.3),

$$
= \frac{2D}{M^2} \exp\left(-2\frac{\gamma}{M} t\right) \int_0^t \exp\left(2\frac{\gamma}{M}\tau\right) d\tau
$$

$$
= \frac{D}{\gamma M}\left[1 - \exp\left(-2\frac{\gamma}{M} t\right)\right]
$$

After a long time, regardless of initial velocity, the particle should come into equilibrium with its surroundings. By this we mean that on the average no transfer of kinetic energy is taking place between the fluid and the particle. On general grounds (the equipartition theorem), the equilibrium distribution of the velocity should be Maxwellian

$$
W(v) = \left(\frac{M}{2\pi kT}\right)^{1/2} \exp\left(\frac{-Mv^2}{2kT}\right) \tag{7.2.7}
$$

where $k = R/N$ is Boltzmann's constant. Letting $t \to \infty$ in (7.2.6) gives the limiting variance $D/\gamma M$ and equating this value with the variance of (7.2.7) yields

$$
D = kT\gamma \tag{7.2.8}
$$

In the next section we use this to show the Einstein relation (7.1.1).

7.3   THE DISPLACEMENT OF THE BROWNIAN PARTICLE

Given $V(0) = v_0$ we can obtain the displacement of the Brownian particle by the integration

$$X(t) - X(0) = \int_0^t V(s) \, ds$$

$$= \int_0^t \left[v_0 \exp\left(-\frac{\gamma}{M} s \right) + \frac{1}{M} \int_0^s \exp\left(-\frac{\gamma}{M}(t - \tau) \right) F(\tau) \, d\tau \right] ds$$

$$= \frac{v_0 M}{\gamma} \left(1 - \exp\left(-\frac{\gamma}{M} t \right) \right)$$

$$+ \frac{1}{M} \int_0^t \int_0^s \exp\left(-\frac{\gamma}{M}(s - \tau) \right) F(\tau) \, d\tau \, ds$$

$$(7.3.1)$$

Thus $X(t) - X(0)$ has (conditional) mean

$$E[X(t) - X(0)] = \frac{v_0 M}{\gamma}\left[1 - \exp\left(-\frac{\gamma}{M} t \right) \right] \tag{7.3.2}$$

A similar series of calculations as in (7.2.6), using the value of D from (7.2.8), gives the (conditional) variance

$$E\left[\left((X(t) - X(0)) - \frac{v_0 M}{\gamma}\left(1 - \exp\left(-\frac{\gamma}{M} t \right) \right) \right)^2 \right]$$

$$= \frac{2kTt}{\gamma} + \frac{kTM}{\gamma^2}\left[-3 + 4\exp\left(-\frac{\gamma}{M} t \right) - \exp\left(-2\frac{\gamma}{M} t \right) \right] \tag{7.3.3}$$

The unconditional mean can be obtained by averaging (7.3.2) with respect to the Maxwellian density (7.2.7),

$$E[X(t) - X(0)] = 0 \tag{7.3.4}$$

Similarly, using (7.3.2) and (7.3.3) and averaging

$$E[(X(t) - X(0))^2]$$

$$= E[((X(t) - X(0)) - E[X(t) - X(0)])^2] + (E[X(t) - X(0)])^2$$

$$= \frac{2kTt}{\gamma} + \frac{kTM}{\gamma^2}\left[-3 + 4\exp\left(-\frac{\gamma}{M} t \right) - \exp\left(-2\frac{\gamma}{M} t \right) \right]$$

$$+ \left[\frac{v_0 M}{\gamma}\left(1 - \exp\left(-\frac{\gamma}{M} t \right) \right) \right]^2$$

with respect to the same density, we get the unconditional variance

$$E[(X(t) - X(0))^2] = \frac{2kT}{\gamma}\left[t - \frac{M}{\gamma}\left(1 - \exp\left(-\frac{\gamma}{M} t \right) \right) \right] \tag{7.3.5}$$

We do not do the calculations here, since they will be verified by a simpler method at the end of the next section. For $t >> M/\gamma$ this says that

$$E[(X(t) - X(0))^2] \sim \frac{2kTt}{\gamma}$$

which is Einstein's formula.

7.4 BROWNIAN MOTION

In order to obtain further information about Brownian motion it is necessary to postulate that $F(t)$, in addition to being stationary with singular covariance, satisfies the rule for higher correlations of Section (5.4). With this assumption $F(t)$ is interpreted as a Gaussian process, and since its spectral density is constant over the whole frequency range

$$\frac{1}{2\pi} \int_{-\infty}^{\infty} e^{-i\omega t} \, 2D\delta(t) \, dt = \frac{D}{\pi}$$

$F(t)$ is referred to as white noise. Of course there really are no processes with singular covariance, but there are processes whose covariance approximates this δ-function covariance as closely as you like. The Ornstein–Uhlenbeck process with

$$r(t) = \frac{D}{\epsilon} \exp\left(-\frac{|t|}{\epsilon}\right)$$

is a Gaussian process which is very nearly of this singular covariance, if ϵ is close to zero. The formal calculations that appear below can be justified by using this process instead of $F(t)$ and then letting $\epsilon \to 0$. With this higher correlation assumption on $F(\tau)$, $V(t)$ is a Gaussian process, since one can show by using (7.2.4) that $V(t)$ also satisfies the rule for higher correlations.

From this and the (conditional) mean and variance of Section (7.2) we then know the conditional density

$P[v, t | v_0, 0]$

$$= \left[\frac{\gamma M}{2\pi D(1 - \exp(-2(\gamma/M)t))}\right]^{1/2} \exp\left[-\frac{\gamma M}{2D} \frac{(v - v_0 \exp(-(\gamma/M)t))^2}{1 - \exp(-2(\gamma/M)t)}\right]$$

$$(7.4.1)$$

which by Einstein's relation, $D = kT\gamma$, is

$$= \left[\frac{M}{2\pi kT(1 - \exp(-2(\gamma/M)t))}\right]^{1/2} \exp\left[-\frac{M}{2kT} \frac{(v - v_0 \exp(-(\gamma/M)t))^2}{1 - \exp(-2(\gamma/M)t)}\right]$$

Averaging this with respect to the Maxwellian distribution (7.2.7), we see that this distribution is stationary for $V(t)$. Further, since it is clear from the definition of $V(t)$ that

$$V(t + s) = V(s) \exp\left(-\frac{\gamma}{M} t\right) + \frac{1}{M} \int_s^{t+s} \exp\left(-\frac{\gamma}{M}(t + s - \tau)\right) F(\tau) \, d\tau$$

(7.4.2)

and $F(t)$ is stationary, we conclude that with this initial distribution $V(t)$ is a stationary Gaussian process. From (7.4.1) we then find the covariance

$$r(t) = \frac{kT}{M} \exp\left(-\frac{\gamma}{M} |t|\right)$$

(7.4.3)

which implies, by Doob's theorem, that $V(t)$ is an Ornstein–Uhlenbeck process (i.e., it is stationary Gaussian Markovian).

Assuming that the Brownian particle has been "kicked around" long enough to have this stationary distribution, we can now easily verify formulas (7.3.4) and (7.3.5) for the (unconditional) mean

$$E[X(t) - X(0)] = \int_0^t E[V(\tau)] \, d\tau = 0$$

and (unconditional) variance

$$E[(X(t) - X(0))^2] = \int_0^t \int_0^t E[V(\eta)V(\tau)] \, d\eta \, d\tau$$

by (7.4.3)

$$= \frac{kT}{M} \int_0^t \int_0^t \exp\left(-\frac{\gamma}{M} |\eta - \tau|\right) d\eta \, d\tau$$

$$= \frac{2kT}{M} \int_0^t \int_0^\tau \exp\left(-\frac{\gamma}{M} (\tau - \eta)\right) d\eta \, d\tau$$

$$= \frac{2kT}{\gamma} \left[t - \int_0^t \exp\left(-\tau \frac{\gamma}{M}\right) \right] d\tau$$

$$= \frac{2kT}{\gamma} \left[t - \frac{M}{\gamma}\left(1 - \exp\left(-t \frac{\gamma}{M}\right)\right) \right]$$

Since $V(t)$ is a Gaussian process subject either to a fixed initial velocity $V(0) = v_0$ or subject to the initial Maxwellian distribution of velocities, we know that the displacement

$$X(t) - X(0) = \int_0^t V(s) \, ds$$

is also a Gaussian process (although not Markovian), subject to either of these same initial distributions of velocities. This follows from the result that an integral of a Gaussian process is also a Gaussian process. Therefore, the (conditional) distribution of $X(t) - X(0)$ is normal of mean (7.3.2) and variance (7.3.3). The unconditional distribution of $X(t) - X(0)$ is normal of mean zero and variance (7.3.5). Here there is no stationary distribution, since as $t \to \infty$ the variance $\to \infty$.

7.5   THE FOKKER–PLANCK EQUATION

For a stochastic process $X(t)$ we may interpret the variation of the density function $W(x, t)$ over time as a macroscopic description of many microscopic particles undergoing motions governed by the process. The Fokker–Planck equation is a differential equation that describes these variations for a particular class of processes. The fundamental solution of this equation turns out to determine the process. Thus the study of certain random processes and their associated Fokker–Planck differential equations are intimately related.

Assume that $X(t)$ is a Markov process with the following properties:

(i) $P(x, s + t | y, s)$ is independent of s, which is referred to as having stationary transition probabilities, and

(ii)   as $\Delta t \to 0$, only the first and second moments of the change in position become proportional to Δt,

(a) $E[X(\Delta t) - X(0) | X(0) = x_0] = A(x_0) \, \Delta t + o(\Delta t)$

(b) $E[(X(\Delta t) - X(0))^2 | X(0) = x_0] = B(x_0) \, \Delta t + o(\Delta t)$ (7.5.1)

(c) $E[(X(\Delta t) - X(0))^k | X(0) = x_0] = o(\Delta t), \quad k > 2$

We can show that the transition densities of $X(t)$ satisfy a differential equation, which is called the Fokker–Planck equation. Define $P(x, t | y) = P(x, s + t | y, s)$, and write the Chapman–Kolmogorov equation (6.3.2) in the form

$$P(x, t + \Delta t | x_0) = \int_{-\infty}^{\infty} P(\xi, t | x_0) P(x, \Delta t | \xi) \, d\xi \qquad (7.5.2)$$

Let ϕ be a smooth function (infinitely differentiable, with bounded derivatives of all orders) so that

$$\phi(x) = \phi(\xi) + (x - \xi)\phi'(\xi) + \tfrac{1}{2}(x - \xi)^2 \phi''(\xi) + \cdots$$

which in addition as $x \to \pm \infty$,

$$\phi(x), \; \phi'(x) \to 0$$

Integrating (7.5.2) against ϕ, we obtain

$$\int_{-\infty}^{\infty} P(x, t + \Delta t | x_0)\phi(x) \, dx = \int_{-\infty}^{\infty} P(\xi, t | x_0) \left(\int_{-\infty}^{\infty} P(x, \Delta t | \xi)\phi(x) \, dx \right) d\xi$$

or, since

$$\int_{-\infty}^{\infty} P(x, \Delta t | \xi)\phi(x) \, dx$$

$$= \int_{-\infty}^{\infty} P(x, \Delta t | \xi)(\phi(\xi) + (x - \xi)\phi'(\xi) + \tfrac{1}{2}(x - \xi)^2\phi''(\xi) + \cdots) \, dx$$

$$= \phi(\xi) + E[X(\Delta t) - X(0) | X(0) = \xi]\phi'(\xi)$$
$$\quad + \tfrac{1}{2}E[(X(\Delta t) - X(0))^2 | X(0) = \xi]\phi''(\xi) + \cdots$$

$$= \phi(\xi) + (A(\xi)\Delta t + o(\Delta t))\phi'(\xi) + \tfrac{1}{2}(B(\xi) + o(\Delta t))\phi''(\xi) + \cdots$$

we have

$$\int_{-\infty}^{\infty} P(x, t + \Delta t | x_0)\phi(x) \, dx = \int_{-\infty}^{\infty} P(\xi, t | x_0)\phi(\xi) \, d\xi$$

$$+ \int_{-\infty}^{\infty} P(\xi, t | x_0)(A(\xi) \, \Delta t + o(\Delta t))\phi'(\xi) \, d\xi$$

$$+ \int_{-\infty}^{\infty} P(\xi, t | x_0)\tfrac{1}{2}(B(\xi) \, \Delta t + o(\Delta t))\phi''(\xi) \, d\xi + \cdots$$

Rearranging, setting $\xi = x$, and letting $\Delta t \to 0$ gives

$$\int_{-\infty}^{\infty} \frac{\partial}{\partial t} (P(x, t | x_0))\phi(x) \, dx$$

$$= \int_{-\infty}^{\infty} P(x, t | x_0)A(x)\phi'(x) \, dx + \int_{-\infty}^{\infty} P(x, t | x_0)\tfrac{1}{2}B(x)\phi''(x) \, dx$$

which upon integration by parts

$$= \int_{-\infty}^{\infty} \frac{-\partial}{\partial x} (P(x, t | x_0)A(x))\phi(x) \, dx + \int_{-\infty}^{\infty} \frac{1}{2} \frac{\partial^2}{\partial x^2} (P(x, t | x_0)B(x))\phi(x) \, dx$$

Since this is true for all functions ϕ, we conclude that

$$\frac{\partial}{\partial t} P(x, t|x_0) = -\frac{\partial}{\partial x} [A(x)P(x, t|x_0)] + \frac{1}{2} \frac{\partial^2}{\partial x^2} [B(x)P(x, t|x_0)] \qquad (7.5.3)$$

which is the Fokker–Planck equation. Notice that it is homogeneous in time, meaning that the transition density

$$P(x, t|y, s)$$

satisfies

$$\frac{\partial}{\partial t} P(x, t|y, s) = -\frac{\partial}{\partial x} [A(x)P(x, t|y, s)] + \frac{1}{2} \frac{\partial^2}{\partial x^2} [B(x)P(x, t|y, s)]$$

This is also called the forward equation, since it involves fixing an initial time s and finding the derivative of the transition density

$$P(x, t|y, s)$$

with respect to the "future" time t.

Similarly, there is an equation, called the backward equation, formed from fixing t and differentiating with respect to the "past" time s. This equation has the form

$$\frac{\partial}{\partial s} P(x, t|y, s) = -A(y) \frac{\partial}{\partial y} P(x, t|y, s) - \frac{1}{2} B(y) \frac{\partial^2}{\partial y^2} P(x, t|y, s)$$

where A and B are as above.

7.6 THE FOKKER–PLANCK EQUATION AND BROWNIAN MOTION

The Wiener process, advanced by Einstein as a reasonable approximation to Brownian motion, satisfies the assumptions necessary for the derivation of the Fokker–Planck equation. The coefficients, given by (7.5.1), are $A(x) = 0$, $B(x) = B$. The corresponding differential equation is

$$\frac{\partial}{\partial t} P(x, t|x_0) = \frac{B}{2} \frac{\partial^2}{\partial x^2} P(x, t|x_0) \qquad (7.6.1)$$

which we solve subject to initial condition $P(x, 0|x_0) = \delta(x - x_0)$ and boundary conditions $P, \partial P/\partial x \to 0$ as $x \to \pm\infty$, by using the Fourier inversion formula.

Taking the Fourier transform of (7.6.1), we obtain

$$\int_{-\infty}^{\infty} \frac{\partial}{\partial t} (P(x,\ t|x_0)) e^{i\xi x}\ dx = \frac{B}{2} \int_{-\infty}^{\infty} \frac{\partial}{\partial x^2} (P(x,\ t|x_0) e^{i\xi x}\ dx$$

and integrating by parts,

$$= \frac{-B\xi^2}{2} \int_{-\infty}^{\infty} e^{i\xi x} P(x,\ t|x_0)\ dx.$$

Defining

$$Q(\xi,\ t) = \int_{-\infty}^{\infty} e^{i\xi x} P(x,\ t|x_0)\ dx \tag{7.6.2}$$

this equation is, formally,

$$\frac{dQ}{dt} = -\tfrac{1}{2} B\xi^2 Q(\xi,\ t)$$

which has the solution

$$Q(\xi,\ t) = K(\xi) e^{-(1/2)B\xi^2 t} \tag{7.6.3}$$

Setting $t = 0$ in (7.6.2) and (7.6.3) and using the initial condition $P(x,\ 0|x_0) = \delta(x - x_0)$ gives

$$K(\xi) = e^{i\xi x_0}$$

which implies that

$$Q(\xi,\ t) = e^{i\xi x_0} e^{-(1/2)B\xi^2 t}$$

Applying the Fourier inversion formula to this, we get

$$P(x,\ t|x_0) = \frac{1}{2\pi} \int_{-\infty}^{\infty} Q(\xi,\ t) e^{-i\xi x}\ d\xi$$

$$= \frac{1}{2\pi} \int_{-\infty}^{\infty} (e^{i\xi x_0} e^{-(1/2)B\xi^2 t}) e^{-i\xi x}\ d\xi$$

and by (1.7.7)

$$= \frac{1}{\sqrt{2\pi B t}} \exp\left[-\frac{(x - x_0)^2}{2Bt} \right]$$

We can also find the coefficients of the Fokker–Planck equation for the Ornstein–Uhlenbeck process $V(t)$. From (7.2.4) we have

$$E[V(\Delta t) - V(0)|V(0) = v_0] = v_0 \left[\exp \left(-\frac{\gamma}{M} \Delta t \right) - 1 \right]$$

which by Taylor expansion

$$= -\frac{\gamma}{M} \Delta t \, v_0 + o(\Delta t)$$

Similarly,

$$E[(V(\Delta t) - V(0))^2|V(0) = v_0]$$

$$= E\left[\left(v_0 \left(\exp \left(-\frac{\gamma}{M} \Delta t \right) - 1 \right) + \frac{1}{M} \int_0^{\Delta t} \exp \left(-\frac{\gamma}{M} (\Delta t - \tau) \right) F(\tau) \, d\tau \right)^2 \right]$$

$$= v_0^2 \left(\exp \left(-\frac{\gamma}{M} \Delta t \right) - 1 \right)^2 + \frac{D}{\gamma M} \left(1 - \exp \left(-2 \frac{\gamma}{M} \Delta t \right) \right)$$

which by Taylor expansion

$$= \frac{2D \, \Delta t}{M^2} + o(\Delta t)$$

Applying the rule for higher correlations to $F(\tau)$, one can show that

$$E[(V(\Delta t) - V(0))^k|V(0) = v_0] = o(\Delta t), \qquad k > 2$$

Thus the Fokker–Planck equation for the velocity $V(t)$ is

$$\frac{\partial}{\partial t} P(v, t|v_0) = \frac{\gamma}{M} \frac{\partial}{\partial v} [vP(v, t|v_0)] + \frac{D}{M^2} \frac{\partial^2}{\partial v^2} [P(v, t|v_0)]$$

$$= \frac{\gamma}{M} \frac{\partial}{\partial v} [vP(v, t|v_0)] + \frac{kT\gamma}{M^2} \frac{\partial^2}{\partial v^2} [P(v, t|v_0)]$$

(7.6.4)

It should be noted that although the displacement $X(t) - X(0)$ described by Ornstein–Uhlenbeck is a Gaussian process with stationary transition probabilities, is not a Markov process. This is because the future position of the Brownian particle depends on its velocity, which in turn depends on its past motion. Therefore, we cannot expect the transition densities to satisfy a Fokker–Planck differential equation. In fact, if we calculate (7.5.1) for the displacement, we get $A(x) = 0$, $B(x) = 0$.

7.7 STOCHASTIC DIFFERENTIAL EQUATIONS

In this chapter we have solved the equation

$$aX'(t) + bX(t) = F(t)$$

where $F(t)$ is white noise and $X(t)$ is the velocity of Brownian motion, subject to the frictional force caused by the motion of the particle through the fluid. This is called a stochastic differential equation, since $F(t)$ is random. We found the mean and variance of the solution, and then showed that it was a Gaussian process. If the initial distribution is the stationary distribution, the process has exponential covariance, so that it is an Ornstein–Uhlenbeck process. In particular it is a Markov process, which says that the present velocity contains all the past information that affects the future distribution of velocities. This explains the erratic behavior of the Brownian particle. One can ask if similar results hold for other stochastic differential equations, which are driven by white noise.

One method for solving more general differential equations is to view the output as depending on the driving function. For linear differential equations, the relation between input and output is a linear system. We have seen that one way of expressing the solution to linear systems is in terms of the impulse response function.

Consider a general linear homogeneous differential equation

$$a_n x^{(n)}(t) + \cdots + a_1 x'(t) + a_0 x(t) = 0$$

We shall assume that this differential equation is stable, which means that for any initial conditions, every solution vanishes at infinity. The impulse response function $f(t)$ is defined as its solution with the initial conditions

$$x^{(n-1)}(0) = \frac{1}{a_n}$$
$$x^{(n-2)}(0) = \cdots = x(0) = 0$$

where $x(t) = 0$ for $t < 0$. If the equation is second order, giving the displacement of a particle, this corresponds to having a particle at rest at the origin, and imparting a unit impulse $\Delta MV = 1$ at $t = 0$. According to the theory of differential equations, if the differential equation is driven by a "reasonable" function $g(t)$,

$$a_n x^{(n)}(t) + \cdots + a_1 x'(t) + a_0 x(t) = g(t)$$

the solution, subject to the initial conditions

$$x(0) = \cdots = x^{(n-1)}(0) = 0$$

is

$$x(t) = \int_0^\infty f(t - s)g(s) \, ds$$

where f is the impulse response function (reasonable means that g is continuous and does not grow too fast as $t \to -\infty$). One can see formally that f is in agreement with our previous discussion of impulse response functions by setting the driving function to be the Dirac delta function $g(s) = \delta(s)$ and obtaining

$$x(t) = \int_0^\infty f(t - s)\delta(s) \, ds = f(t)$$

If the initial conditions are not all zero, one introduces the solutions $\phi_k(t)$ to the homogeneous equation

$$a_n x^{(n)}(t) + \cdots + a_1 x'(t) + a_0 x(t) = 0$$

subject to the initial conditions

$$x(0) = \cdots = x^{(k-1)}(0) = 0, \, x^{(k)}(0) = 1, \, x^{(k+1)}(0) = \cdots = x^{(n-1)}(0) = 0$$

and the general solution, now with initial conditions

$$x(0) = b_1, \ldots, x^{(n-1)}(0) = b_{n-1}$$

and driving function g is

$$x(t) = b_1 \phi_1(t) + \cdots + b_{n-1}\phi_{n-1}(t) + \int_0^\infty f(t - s)g(s) \, ds$$

In the stochastic case,

$$a_n X^{(n)}(t) + \cdots + a_1 X'(t) + a_0 X(t) = F(t)$$

if the system is driven by white noise $F(t)$, the solution is also in this form

$$X(t) = \int_0^\infty f(t - s)F(s) \, ds$$

This can easily be seen to be the case for the Langevin equation. In case the initial conditions are nonzero, the general solution takes the form

$$X(t) = B_1 \phi_1(t) + \cdots + B_{n-1}\phi_{n-1}(t) + \int_0^\infty f(t - s)F(s) \, ds$$

where

$$B_k = X^{(k)}(t_0)$$

are the initial random variables. From this, we can determine the mean and variance of $X(t)$.

We can use this approach to examine the velocity of a pendulum in a fluid. Here the stochastic differential equation looks something like the Langevin equation:

$$M \frac{d^2X(t)}{dt^2} + \gamma \frac{dX(t)}{dt} + \alpha X(t) = F(t)$$

Here M is the mass of the pendulum, γ the friction coefficient, and α a constant that depends on M and the length of the pendulum.

The impulse response function $f(t)$ is the solution to the homogeneous equation

$$M \frac{d^2f(t)}{dt^2} + \gamma \frac{df(t)}{dt} + \alpha f(t) = 0$$

with the initial conditions

$$\frac{df(0)}{dt} = \frac{1}{M}, \, f(0) = 0$$

which are well known to be of the form

$$f(t) = C(e^{-At} - e^{-Bt})$$

in the overdamped case, and

$$f(t) = Ce^{-At} \sin Bt$$

in the underdamped case, and

$$f(t) = Cte^{-At}$$

in the critical case. Overdamping corresponds to the situation where the fluid is so viscous that once displaced from center the pendulum returns to center without ever crossing the center position. Underdamping corresponds to a low enough viscosity so that the pendulum may perform periodic motion of decreasing amplitude. The critical case corresponds to the viscosity where the behavior changes from underdamped to overdamped. In each case, the solution to the differential equation is given by the integral of f and white noise.

Let us return to the general situation. If we wait long enough, so that the process has come to equilibrium, the solution will be a stationary Gaussian process, and not depend on the initial conditions. The stationary solution can be written

$$X(t) = \int_{-\infty}^{\infty} f(t - s)F(s) \, ds$$

and the corresponding covariance function found from

$$E[X(t)X(0)] = E\left[\int_{-\infty}^{\infty} f(t - s)F(s) \, ds \int_{-\infty}^{\infty} f(-s)F(s) \, ds \right]$$

which is

$$r(t) = \int_{-\infty}^{\infty} \left(\int_{-\infty}^{\infty} f(-s_1)f(t - s_2)\delta(s_1 - s_2) \, ds_2 \right) ds_1$$

or

$$r(t) = \int_{-\infty}^{\infty} f(-s_1)f(t - s_1) \, ds_1$$

Recalling that the frequency response function is

$$A(\omega) = \int_{-\infty}^{\infty} e^{i\omega t} f(t) \, dt$$

and applying the Fourier inversion formula, the spectral density is

$$
\begin{aligned}
f_X(\omega) &= \frac{1}{2\pi} \int_{-\infty}^{\infty} e^{-i\omega t} \int_{-\infty}^{\infty} f(-s_1)f(t - s_1) \, ds_1 \, dt \\
&= \frac{1}{2\pi} \int_{-\infty}^{\infty} e^{-i\omega s_1} f(-s_1) \, ds_1 \int_{-\infty}^{\infty} e^{-i\omega(t - s_1)} f(t - s_1) \, dt \\
&= \frac{1}{2\pi} A(\omega)\overline{A(\omega)} \\
&= \frac{1}{2\pi} |A(\omega)|^2
\end{aligned}
$$

This confirms what we should have suspected, from the relation between the spectral density of input and output of a linear system

$$f_Y(\omega) = |A(\omega)|^2 f_X(\omega)$$

since the a white noise input has constant spectral density. Comparing this with the shot noise example of chapter 4, we see that in fact shot noise is also the output of a linear system driven by white noise.

Finding the frequency response function and hence the spectral density function for the stationary solution to an arbitrary differential equation driven by white noise turns out to be remarkably easy. The equation

$$L(e^{-i\omega t}) = A(\omega)e^{-i\omega t}$$

becomes

$$e^{-i\omega t} = A(\omega)e^{-i\omega t}(a_n(-i\omega)^n + \cdots + a_1(-i\omega) + a_0)$$

so that

$$A(\omega) = \frac{1}{a_n(-i\omega)^n + \cdots + a_0}$$

This explains why one ought to be interested in processes corresponding to rational spectral densities.

PROBLEMS

7.1. Why do Einstein's assumptions imply that the Brownian motion he described is a Markov process?

7.2. Suppose that $X(t)$ is given by the integral

$$X(t) = \int_0^t e^{\alpha(t-s)}F(s) \, ds$$

where α is an arbitrary constant, and $F(t)$ is the white noise process.

(a) Show that $X(t)$ is a Gaussian process.

(b) Find the mean and covariance functions for this process.

7.3. Let $X(t)$ be given by the integral

$$X(t) = \int_{-\infty}^t e^{\alpha(t-s)}F(s) \, ds$$

where $\alpha < 0$ and $F(t)$ is the white noise process.

(a) Show that $X(t)$ is a stationary Gaussian process.

(b) Find its covariance function.

(c) Find the stationary distribution.

7.4. Show that the formal solution to the differential equation (7.2.2) is of the form (7.2.4).

7.5. Show that (7.2.4) obeys the rule for higher correlations, and thus conclude that it is a normal random variable.

7.6. Verify that the Maxwellian distribution (7.2.7) is stationary for $V(t)$ given in (7.2.4).

7.7. Show that subject to an initial Maxwellian distribution of velocities, the displacement $X(t) - X(0)$ is a stationary Gaussian process, but not Markovian. Use the generalization of Doob's theorem to show that this is true with a fixed initial velocity as well.

7.8. Suppose that $F(t)$ is white noise, and

$$Z(t) = \int_0^t F(s)\, ds$$

(a) Show that $Z(t)$ is Gaussian, by demonstrating the rule for higher correlations.

(b) Show that $Z(t)$ has independent identically distributed increments $Z(t + \Delta t) - Z(t)$.

(c) Calculate the covariance function of $Z(t)$.

(d) Explain why we can interpret white noise as the derivative of the Wiener process.

7.9. Suppose that $N(t)$ is a Poisson process and

$$V(t) = (-1)^{N(t)}$$

We can interpret this as an approximation to the one-dimensional velocity of the Brownian particle, and approximate the displacement by

$$X(t) = \int_0^t (-1)^{N(t)}\, dt$$

(a) Find the mean and covariance functions for this process.

(b) Show that this matches our model, in that as $t \to \infty$ the variance is proportional to t, and as $t \to 0$ the variance is proportional to t^2.

7.10. Let $W(t)$ be the Wiener process and $a(t)$ a differentiable function. Show that the "integration by parts formula"

$$\int_0^t a(s)\, dW(s) = a(s)W(s)\Big|_0^t - \int_0^t a'(s)W(s)\, ds$$

defines the same Gaussian process as our interpretation of

$$\int_0^t a(s)F(s)\, ds$$

by calculating the mean and covariance function for each of these processes. This gives a second interpretation of white noise as the derivative of the Wiener process.

7.11. Under general conditions, we have shown that the transition density of a Markov process $P(x, t|y, s)$ satisfies the Fokker–Planck equation

$$\frac{\partial}{\partial t} P(x, t|y, s) = -\frac{\partial}{\partial x} [A(x)P(x, t|y, s)] + \frac{1}{2}\frac{\partial^2}{\partial x^2} [B(x)P(x, t|y, s]$$

This equation involves differentiation with respect to future times and is often referred to as the forward equation. Similarly, there is a differential equation which involves the past, called the backward equation,

$$\frac{\partial}{\partial s} P(x, t|y, s) = -A(x)\frac{\partial}{\partial y} P(x, t|y, s) + \frac{1}{2}B(x)\frac{\partial^2}{\partial y^2} P(x, t|y, s)$$

where the coefficients, $A(x)$ and $B(x)$ have the same definitions as in the forward equation. Show that the transition densities for both the Wiener process and the Ornstein–Uhlenbeck process satisfy both their forward and backward equations.

7.12. How big should the impulses $\pm\Delta(MV)$, mentioned in Section 7.2, be if one is made in every time interval Δt, in order that the variance of the resulting velocity at time $t = n\Delta t$ matches the (conditional) variance of $V(t)$?

7.13. Find the formal spectral density of white noise.

7.14. Find the spectral density of the process, with approximately singular covariance, given in Section 7.4.

7.15. Show directly that (7.3.5) holds.

7.16. Suppose that $A(x) = a$ and $B(x) = b$ are constants in the Fokker–Planck equation. Show that the associated Markov process can be written

$$at + bW(t)$$

where $W(t)$ is the Wiener process. Therefore, a can be interpreted as the velocity of a drift imposed on the Wiener process.

7.17. Show that $(X(t), V(t))$ is a Markov process.

CHAPTER
8

Markov Chains

In Chapter 7 we discussed the nature of Brownian motion based on a few simple assumptions, using a differential equation from physics. Another approach towards understanding this phenomenon is to examine certain discrete time, discrete space, Markov processes, and view the physical process as a limiting case. Discrete time, discrete space, Markov processes are called Markov chains. They have broad applications and a large literature associated with them. We shall content ourselves with showing a few examples related to the theory of Brownian motion, beginning with an informal discussion of the Ehrenfest model.

8.1 THE EHRENFEST DOG-FLEA MODEL

Perhaps the best known and most instructive example of a Markov chain is the Ehrenfest "dog-flea" model proposed in 1907 by P. and T. Ehrenfest to demonstrate a probabilistic approach to the physical notion of equilibrium. Imagine two boxes (dogs) in which $2R$ balls (fleas) numbered $1, 2, \ldots, 2R$ are distributed. We choose at random an integer between 1 and $2R$, move the ball with that number from the box it is in to the other, and repeat this procedure indefinitely. Denoting by $X(n)$ the number of balls in the first box after n repetitions, $X(n)$ is a Markov chain, satisfying

$$P[X(n + 1) = j | X(n) = i] = \begin{cases} (2R - i)/2R, & j = i + 1 \\ i/2R, & j = i - 1 \\ 0, & \text{else} \end{cases} \quad (8.1.1)$$

The probabilities above are called the transition probabilities of the Markov chain and are usually denoted p_{ij}:

$$p_{ij} = P[X(n + 1) = j | X(n) = i] \tag{8.1.2}$$

which is independent of n. The matrix of transition probabilities $\mathbf{P} = (p_{ij})$ is called the transition matrix of the Markov chain. If the initial state is random, with probability mass function $w(k) = P[X(0) = k]$, $k = 0, 1, \ldots, 2R$, it is easy to see that the distribution of $X(1)$ is given by

$$P[X(1) = i] = \sum_k w(k)p_{ki} \tag{8.1.3}$$

which corresponds to the entries in the matrix product $\mathbf{wP}$, where $\mathbf{w} = (w(0), \ldots, w(2R))$. Similarly, after s repetitions, the entries of the product $\mathbf{wP}^s$ give the distribution of $X(s)$. Thus the study of this process can be reduced to the study of the matrix $\mathbf{P}$ which by (8.1.1) is of the form

$$\mathbf{P} = \begin{bmatrix} 0 & 1 & 0 & 0 & \cdots & 0 & 0 & 0 & 0 \\ 1/2R & 0 & 1 - 1/2R & 0 & \cdots & 0 & 0 & 0 & 0 \\ 0 & 2/2R & 0 & 1 - 2/2R & \cdots & 0 & 0 & 0 & 0 \\ \vdots & \vdots & \vdots & \vdots & \vdots & \vdots & \vdots & \vdots & \vdots \\ 0 & 0 & 0 & 0 & \cdots & 1 - 2/2R & 0 & 2/2R & 0 \\ 0 & 0 & 0 & 0 & \cdots & 0 & 1 - 1/2R & 0 & 1/2R \\ 0 & 0 & 0 & 0 & \cdots & 0 & 0 & 1 & 0 \end{bmatrix}$$

Further, it is not hard to show that for the Ehrenfest model the initial distribution $\pi(k)$ given by

$$\pi(k) = \frac{(2R)!}{(k)!(2R - k)!} (1/2)^{2R} \tag{8.1.4}$$

is stationary in the sense that

$$\sum_k \pi(k)p_{ki} = \pi(i) \tag{8.1.5}$$

or, equivalently,

$$\pi\mathbf{P} = \pi$$

and with this initial distribution $X(n)$ becomes a stationary process.

One can also check that with this stationary initial distribution the model is "time reversible" meaning that

$$P[X(n + 1) = j | X(n) = i] = P[X(n - 1) = j | X(n) = i] \tag{8.1.6}$$

To show this we go back to the definition of conditional probabilities,

$$P[X(n - 1) = j | X(n) = i] = \frac{P[X(n - 1) = j, X(n) = i]}{P[X(n) = i]}$$

$$= \frac{P[X(n) = i | X(n - 1) = j] P[X(n - 1) = j]}{P[X(n) = i]}$$

$$= \frac{p_{ji} \pi(j)}{\pi(i)}$$

so (8.1.6) is equivalent to

$$\pi(i) p_{ij} = \pi(j) p_{ji} \tag{8.1.7}$$

This can be easily verified by using the formulas (8.1.1) and (8.1.4). It was this property that made the model interesting as a model for heat exchange in physical systems.

If we also introduce the conditional probabilities

$$p_{ij}(s) = P[X(n + s) = j | X(n) = i] \tag{8.1.8}$$

which are referred to as the s-step transition probabilities, then since $X(n)$ is Markov, one can see that

$$p_{ij}(s + 1) = p_{j-1j} p_{ij-1}(s) + p_{j+1j} p_{ij+1}(s)$$

We shall show later in the chapter, by an appropriate limiting process, that this equation becomes the Fokker–Planck equation of the Ornstein–Uhlenbeck process. Thus the Ehrenfest model may be interpreted as a discrete approximation of this important process. The Wiener process is similarly related to another simple Markov chain, as we shall see in Section 8.6.

8.2   MARKOV CHAINS

We now give the formal definition of a Markov chain. Let $X(n) = X_n$, $n = 0, 1, 2,$ . . . be a family of random variables, taking values from a, not necessarily numeric, collection of symbols or "states" $\{S_i, i = 1, 2, \ldots\}$, and define p_{ij} by

$$p_{ij} = P[X_{n+1} = S_j | X_n = S_i] \tag{8.2.1}$$

and assume this is independent of n (i.e., the transition probabilities are stationary). Viewing X_n as a stochastic process it will be Markovian if

$$P[X_n = S_{i_n} | X_{n-1} = S_{i_{n-1}}, \ldots, X_1 = S_{i_1}] = P[X_n = S_{i_n} | X_{n-1} = S_{i_{n-1}}]$$

which implies

$$P[X_0 = S_{i_0}, X_1 = S_{i_1}, \ldots, X_n = S_{i_n}] \tag{8.2.2}$$
$$= w[S_{i_0}]P[X_1 = S_{i_1}|X_0 = S_{i_0}] \ldots P[X_n = S_{i_n}|X_{n-1} = S_{i_{n-1}}]$$

where $w[S_k] = P[X_0 = S_k]$, $k = 1, 2, \ldots$, is an arbitrarily prescribed initial distribution. Such a process is called a Markov chain.

Adopting the notation

$$p_{ij}(s) = P[X_{n+s} = S_j|X_n = S_i] \tag{8.2.3}$$

the Markov property (8.2.2) and Komogorov consistency implies that

$$p_{ij}(2) = \sum_k p_{ik} p_{kj} \tag{8.2.4}$$

which is a special case of the Chapman-Kolmogorov equation (6.3.2). Assuming that the number of states is finite, and letting $\mathbf{P} = (p_{ij})$ be the matrix of transition probabilities, these are the entries of the matrix $\mathbf{P}^2$,

$$\mathbf{P}^2 = (p_{ij}(2))$$

In general we have

$$\mathbf{P}^s = (p_{ij}(s)) \tag{8.2.5}$$

so that given an initial distribution $w(k) = P[X_0 = S_k]$, the distribution of X_s is given by the entries of the matrix $\mathbf{w}\mathbf{P}^s$.

From this, it is clear that all the pertinent information about a finite state Markov chain, except its initial distribution, is contained in the matrix $\mathbf{P}$, whose entries satisfy

$$\sum_j p_{ij} = 1 \tag{8.2.6}$$

for each i, and

$$p_{ij} \geq 0 \tag{8.2.7}$$

Any matrix $\mathbf{P}$ satisfying these two conditions is called a stochastic matrix, and corresponding to any stochastic matrix is a Markov chain. Thus the theory of finite-state Markov chains is in fact the theory of these matrices. The probabilistic flavor of this theory comes from examples, as is true for other parts of the theory of stochastic processes.

8.3 PROPERTIES OF STOCHASTIC MATRICES

In this section we investigate some of the properties of stochastic matrices, using basic concepts from linear algebra. First we make two definitions. Suppose that $\mathbf{P}$ is a stochastic matrix. If

$$\lim_{s \to \infty} \mathbf{P}^s = \mathbf{H} \qquad (8.3.1)$$

exists, we call $\mathbf{H}$ the **limiting distribution** of the Markov chain. If $\mathbf{H}$ is a matrix of identical rows, then any row of $\mathbf{H}$ is called the **equilibrium distribution** of the Markov chain. The reasoning behind these definitions is as follows. The ith row of $\mathbf{P}^s$ lists the conditional probabilities

$$P[X_s = S_j | X_0 = S_i]$$

If there is a limit, the distribution of X_s approaches a fixed distribution as $s \to \infty$, depending on the initial distribution. If $\mathbf{H}$ has identical rows, this distribution does not depend on the initial distribution, and deserves the name "equilibrium" distribution. We note that if $\mathbf{P}$ has a limiting distribution, the relation

$$\mathbf{HP} = \mathbf{H} \qquad (8.3.2)$$

implies that the rows of $\mathbf{H}$ satisfy the stationary relation (8.1.5).

The matrix $\mathbf{P}$ has 1 as an eigenvalue, since

$$\mathbf{P}\begin{bmatrix} 1 \\ 1 \\ \vdots \\ 1 \end{bmatrix} = \begin{bmatrix} 1 \\ 1 \\ \vdots \\ 1 \end{bmatrix} \qquad (8.3.3)$$

All the rest of its eigenvalues are less than or equal to 1 in absolute value. To show this, let λ be an eigenvalue of largest absolute value. Since λ is an eigenvalue there exists $\mathbf{x} = (x_1, \ldots, x_n)^\mathrm{T}$ such that

$$\mathbf{Px} = \lambda \mathbf{x}$$

so for each coordinate x_i

$$\sum_{j=1}^{n} p_{ij} x_j = \lambda x_i$$

Therefore, for any i,

$$\sum_{j=1}^{n} p_{ij}|x_j| \geq |\lambda| \, |x_i|$$

and if $|x_k|$ is the largest of the $|x_i|$'s,

$$\sum_{j=1}^{n} p_{kj}|x_k| \geq \sum_{j=1}^{n} p_{kj}|x_i| \geq |\lambda| \, |x_k|$$

so that

$$\sum_{j=1}^{n} p_{kj} \geq |\lambda|$$

which is

$$|\lambda| \leq 1 \tag{8.3.4}$$

The limiting distribution of a stochastic matrix can be easily found if the matrix is diagonalizable, that is,

$$\mathbf{P} = \mathbf{C}^{-1}\mathbf{D}\mathbf{C}$$

where $\mathbf{D}$ is a diagonal matrix. The diagonal entries d_{ii} of $\mathbf{D}$ are of course eigenvalues of $\mathbf{P}$. Since

$$\mathbf{P}^s = \mathbf{C}^{-1}\mathbf{D}^s\mathbf{C}$$

and $\mathbf{D}^s$ is a diagonal matrix with diagonal entries $(d_{ii})^s$, a diagonalizable stochastic matrix has a limiting distribution if and only if the only eigenvalues with $|\lambda| = 1$ are $\lambda = 1$ and in this case

$$\lim_{s \to \infty} \mathbf{P}^s = \mathbf{C}^{-1}\mathbf{E}\mathbf{C} \tag{8.3.5}$$

where E is a diagonal matrix with $e_{ii} = 0$ if $d_{ii} < 1$, $e_{ii} = 1$ if $d_{ii} = 1$.

We now give a short discussion of diagonalization. Suppose that $\mathbf{P}$ is diagonalizable

$$\mathbf{P} = \mathbf{C}^{-1}\mathbf{D}\mathbf{C} \tag{8.3.6}$$

Writing this as

$$\mathbf{P}\mathbf{C}^{-1} = \mathbf{C}^{-1}\mathbf{D}$$

we see that (8.3.6) will hold iff there is an invertible matrix, whose columns are right eigenvectors of $\mathbf{P}$, which is the matrix $\mathbf{C}^{-1}$. Similarly, it will hold if there is an invertible matrix whose rows are left eigenvectors, which is the matrix $\mathbf{C}$. Since left and right eigenvectors corresponding to different eigenvalues are orthogonal, once having chosen $\mathbf{C}$ to be a matrix of left eigenvectors, one can use this to find $\mathbf{C}^{-1}$ by making a selection of right eigenvectors subject only to the condition that left and right eigenvectors corresponding to the same eigenvalue multiply correctly. Related to this we have the following theorem.

Suppose that $\mathbf{P}$ is diagonalizable and its only eigenvalue with $|\lambda| = 1$ is $\lambda = 1$. If there exists a vector $\mathbf{v} = (v_1, \ldots, v_m)$ with $\Sigma\, v_i = 1$ such that every left eigenvector corresponding to the eigenvalue 1 can be written $\alpha\mathbf{v}$ with α real, then $\mathbf{P}$ has an equilibrium distribution equal to $\mathbf{v}$.

To show this consider the diagonalization of $\mathbf{P}$,

$$\mathbf{P} = \mathbf{C}^{-1}\mathbf{DC}$$

By what we have stated above we may choose $\mathbf{C}^{-1}$ to have as its first column a column of 1's, $\mathbf{C}$ to have as its first row $(v_1, \ldots, v_n)$, and $\mathbf{D}$ to have first diagonal entry 1 and all other diagonal entries of absolute value less than 1. Therefore,

$$\lim_{s\to\infty} \mathbf{P}^s = \lim_{s\to\infty} \mathbf{C}^{-1}\mathbf{D}^s\mathbf{C}$$

$$= \mathbf{C}^{-1}\mathbf{EC}$$

where $\mathbf{E}$ is a matrix of zeros except for a 1 in the upper left corner, so

$$\lim_{s\to\infty} \mathbf{P}^s = \begin{bmatrix} v_1 & v_2 & \cdots & v_m \\ \cdot & \cdot & \cdots & \cdot \\ v_1 & v_2 & \cdots & v_m \end{bmatrix} \tag{8.3.7}$$

as required.

As an example, consider a Markov chain with three states and stochastic matrix

$$\mathbf{P} = \begin{bmatrix} 0 & 1/2 & 1/2 \\ 1/2 & 0 & 1/2 \\ 1/2 & 1/2 & 0 \end{bmatrix}$$

indicating that at each step, the process jumps from the state it is in to one of the other two states, each with probability 1/2. The eigenvalues λ are the roots of the characteristic polynomial

$$|\mathbf{P} - \lambda\mathbf{I}| = 0$$

or

$$(1 - \lambda)(\lambda + 1/2)^2 = 0$$

which gives $\lambda = 1$ and $\lambda = -1/2$. The left eigenvectors corresponding to $\lambda = 1$ are found by solving

$$\mathbf{x}(\mathbf{P} - \mathbf{I}) = 0$$

yielding the one-dimensional solution space: $\mathbf{x} = a(1, 1, 1)$. Thus by the theorem the equilibrium distribution is

$$(1/3, 1/3, 1/3)$$

8.4   APPLICATIONS OF DIAGONALIZATION

If $\mathbf{P}$ is symmetric, it is orthogonally diagonalizable, so we can find an orthonormal set of left eigenvectors. Arranging them as rows in the matrix $\mathbf{C}$, we conclude that

$$\mathbf{P} = \mathbf{C}^{-1}\mathbf{D}\mathbf{C}$$

where

$$\mathbf{C}^{-1} = \mathbf{C}^{\mathrm{T}}$$

Recall that eigenvectors associated with different eigenvalues are automatically orthogonal for symmetric matrices. Therefore, finding orthonormal eigenvectors will involve the Gram–Schmidt orthogonalization process only for the eigenvectors associated with the same eigenvalue.

As an example, let us show the diagonalization of the previous example. We already know from the preceding section that a left normalized eigenvector corresponding to $\lambda = 1$ is $(1/\sqrt{3}, 1/\sqrt{3}, 1/\sqrt{3})$. The left eigenvectors corresponding to $\lambda = -1/2$ are found from

$$\mathbf{x}(\mathbf{P} + \tfrac{1}{2}\mathbf{I}) = 0$$

or

$$x_1 + x_2 + x_3 = 0$$

which has solution

$$\mathbf{x} = \mathbf{a}(-1, 1, 0) + b(-1, 0, 1)$$

Applying the Gram–Schmidt process to $(-1, 1, 0)$ and $(-1, 0, 1)$ yields the orthonormal eigenvectors $(-1/\sqrt{2}, 1/\sqrt{2}, 0)$ and $(1/\sqrt{6}, 1/\sqrt{6}, -2/\sqrt{6})$, so that $\mathbf{C}$ is

$$C = \begin{bmatrix} 1/\sqrt{3} & 1/\sqrt{3} & 1/\sqrt{3} \\ -1/\sqrt{2} & 1/\sqrt{2} & 0 \\ 1/\sqrt{6} & 1/\sqrt{6} & -2/\sqrt{6} \end{bmatrix}$$

We can write the diagonalization $P = C^{-1}DC$ explicitly:

$$\begin{bmatrix} 1/\sqrt{3} & -1/\sqrt{2} & 1/\sqrt{6} \\ 1/\sqrt{3} & 1/\sqrt{2} & 1/\sqrt{6} \\ 1/\sqrt{3} & 0 & -2/\sqrt{6} \end{bmatrix} \begin{bmatrix} 1 & 0 & 0 \\ 0 & -1/2 & 0 \\ 0 & 0 & -1/2 \end{bmatrix} \begin{bmatrix} 1/\sqrt{3} & 1/\sqrt{3} & 1/\sqrt{3} \\ -1/\sqrt{2} & 1/\sqrt{2} & 0 \\ 1/\sqrt{6} & 1/\sqrt{6} & -2/\sqrt{6} \end{bmatrix}$$

and compute P^s

$$\begin{bmatrix} 1/\sqrt{3} & -1/\sqrt{2} & 1/\sqrt{6} \\ 1/\sqrt{3} & 1/\sqrt{2} & 1/\sqrt{6} \\ 1/\sqrt{3} & 0 & -2/\sqrt{6} \end{bmatrix} \begin{bmatrix} 1 & 0 & 0 \\ 0 & (-1/2)^s & 0 \\ 0 & 0 & (-1/2)^s \end{bmatrix} \begin{bmatrix} 1/\sqrt{3} & 1/\sqrt{3} & 1/\sqrt{3} \\ -1/\sqrt{2} & 1/\sqrt{2} & 0 \\ 1/\sqrt{6} & 1/\sqrt{6} & -2/\sqrt{6} \end{bmatrix}$$

which in the limit as $s \to \infty$ becomes

$$\begin{bmatrix} 1/\sqrt{3} & -1/\sqrt{2} & 1/\sqrt{6} \\ 1/\sqrt{3} & 1/\sqrt{2} & 1/\sqrt{6} \\ 1/\sqrt{3} & 0 & -2/\sqrt{6} \end{bmatrix} \begin{bmatrix} 1 & 0 & 0 \\ 0 & 0 & 0 \\ 0 & 0 & 0 \end{bmatrix} \begin{bmatrix} 1/\sqrt{3} & 1/\sqrt{3} & 1/\sqrt{3} \\ -1/\sqrt{2} & 1/\sqrt{12} & 0 \\ 1/\sqrt{6} & 1/\sqrt{6} & -2/\sqrt{6} \end{bmatrix}$$

$$= \begin{bmatrix} 1/3 & 1/3 & 1/3 \\ 1/3 & 1/3 & 1/3 \\ 1/3 & 1/3 & 1/3 \end{bmatrix} \qquad (8.4.1)$$

as stated by the theorem.

Suppose that P is not symmetric but has simple eigenvalues (i.e., the characteristic polynomial factors into distinct linear factors). Then P is diagonalizable, and to find C and C^{-1} we need only find a complete set of right and left eigenvectors and scale them so that right and left eigenvectors corresponding to the same eigenvalue have product 1.

As an example, consider a Markov chain with three states, having stochastic matrix

$$P = \begin{bmatrix} 1 & 0 & 0 \\ 1/2 & 0 & 1/2 \\ 1/2 & 1/2 & 0 \end{bmatrix}$$

indicating that once the process reaches state 1 it stays there, and from state 2 or 3 it jumps to one of the other states each with probability $1/2$. The eigenvalues are the roots of the characteristic polynomial,

$$(1 - \lambda)(\lambda - 1/2)(\lambda + 1/2) = 0$$

which gives $\lambda = 1$, $\lambda = 1/2$, and $\lambda = -1/2$. The left eigenvectors are

$$
\begin{array}{ll}
a(1, 0, 0), & \lambda = 1 \\
b(-2, 1, 1), & \lambda = 1/2 \\
c(0, 1, -1), & \lambda = -1/2
\end{array}
$$

The right eigenvectors are

$$
\begin{array}{ll}
d(1, 1, 1)^{\mathrm{T}}, & \lambda = 1 \\
e(0, 1, 1)^{\mathrm{T}}, & \lambda = 1/2 \\
f(0, 1, -1)^{\mathrm{T}}, & \lambda = -1/2
\end{array}
$$

Fixing the left eigenvectors with $a = b = c = 1$ and scaling the right eigenvectors, so that the product is 1 gives $d = 1$, $e = 1/2$, $f = 1/2$. Thus the diagonalization is

$$
\mathbf{P} =
\begin{bmatrix}
1 & 0 & 0 \\
1 & 1/2 & 1/2 \\
1 & 1/2 & -1/2
\end{bmatrix}
\begin{bmatrix}
1 & 0 & 0 \\
0 & 1/2 & 0 \\
0 & 0 & -1/2
\end{bmatrix}
\begin{bmatrix}
1 & 0 & 0 \\
-2 & 1 & 1 \\
0 & 1 & -1
\end{bmatrix}
$$

from which we can compute

$$
\mathbf{P}^s =
\begin{bmatrix}
1 & 0 & 0 \\
1 & 1/2 & 1/2 \\
1 & 1/2 & -1/2
\end{bmatrix}
\begin{bmatrix}
1 & 0 & 0 \\
0 & (1/2)^s & 0 \\
0 & 0 & (-1/2)^s
\end{bmatrix}
\begin{bmatrix}
1 & 0 & 0 \\
-2 & 1 & 1 \\
0 & 1 & -1
\end{bmatrix}
$$

and

$$
\begin{aligned}
\lim_{s \to \infty} \mathbf{P}^s &=
\begin{bmatrix}
1 & 0 & 0 \\
1 & 1/2 & 1/2 \\
1 & 1/2 & -1/2
\end{bmatrix}
\begin{bmatrix}
1 & 0 & 0 \\
0 & 0 & 0 \\
0 & 0 & 0
\end{bmatrix}
\begin{bmatrix}
1 & 0 & 0 \\
-2 & 1 & 1 \\
0 & 1 & -1
\end{bmatrix} \\
&=
\begin{bmatrix}
1 & 0 & 0 \\
1 & 0 & 0 \\
1 & 0 & 0
\end{bmatrix}
\end{aligned}
\tag{8.4.2}
$$

which once again was predicted by the theorem, since the only eigenvalue with $|\lambda| = 1$ is $\lambda = 1$, and the corresponding left eigenvectors are $a(1, 0, 0)$.

Suppose that $\mathbf{P}$ is diagonalizable and does not satisfy either of the conditions above. Then once $\mathbf{C}$ has been chosen, we can use the fact that left and right eigenvectors corresponding to different eigenvalues are orthogonal to find $\mathbf{C}^{-1}$. As an example, consider a Markov chain with four states and the corresponding stochastic matrix

$$P = \begin{bmatrix} 1 & 0 & 0 & 0 \\ 1/2 & 0 & 1/2 & 0 \\ 0 & 1/2 & 0 & 1/2 \\ 0 & 0 & 0 & 1 \end{bmatrix}$$

indicating that states 1 and 4 are "absorbing" states and at the intermediate states the process jumps to the next larger or smaller state each with probability $1/2$.

The roots of the characteristic polynomial are $\lambda = 1$, $\lambda = 1/2$, $\lambda = -1/2$. The corresponding left eigenvectors are

$$a(1, 0, 0, 0) + b(0, 0, 0, 1), \qquad \lambda = 1$$

$$c(-1, 1, 1, -1), \qquad \lambda = 1/2$$

$$d(-1/3, 1, -1, 1/3), \qquad \lambda = -1/2$$

The right eigenvectors are

$$e(3, 2, 1, 0)^T + f(-2, -1, 0, 1)^T, \qquad \lambda = 1$$

$$g(0, 1, 1, 0)^T, \qquad \lambda = 1/2$$

$$h(0, 1, -1, 0)^T, \qquad \lambda = -1/2$$

Letting C be the matrix of left eigenvectors

$$C = \begin{bmatrix} 1 & 0 & 0 & 0 \\ 0 & 0 & 0 & 1 \\ -1 & 1 & 1 & -1 \\ -1/3 & 1 & -1 & 1/3 \end{bmatrix}$$

we find the first column of C^{-1} by finding numbers e and f satisfying the conditions

$$<e(3, 2, 1, 0)^T + f(-2, -1, 0, 1)^T, (1, 0, 0, 0)> = 1$$

$$<e(3, 2, 1, 0)^T + f(-2, -1, 0, 1)^T, (0, 0, 0, 1)> = 0$$

which gives $e = 1/3$, $f = 0$. So the first column is

$$1/3(3, 2, 1, 0)^T = (1, 2/3, 1/3, 0)^T$$

To find the second column, we solve the system

$$<e(3, 2, 1, 0)^T + f(-2, -1, 0, 1)^T, (1, 0, 0, 0)> = 0$$

$$<e(3, 2, 1, 0)^T + f(-2, -1, 0, 1)^T, (0, 0, 0, 1)> = 1$$

yielding $e = 2/3$, $f = 1$. Therefore, the second column is

$$2/3(3, 2, 1, 0)^T + (-2, -1, 0, 1)^T = (0, 1/3, 2/3, 1)^T$$

so that $\mathbf{C}^{-1}$ is

$$\mathbf{C}^{-1} = \begin{bmatrix} 1 & 0 & 0 & 0 \\ 2/3 & 1/3 & 1/2 & 1/2 \\ 1/3 & 2/3 & 1/2 & -1/2 \\ 0 & 1 & 0 & 0 \end{bmatrix}$$

Using this, the diagonalization of $\mathbf{P}$ is

$$\begin{bmatrix} 1 & 0 & 0 & 0 \\ 2/3 & 1/3 & 1/2 & 1/2 \\ 1/3 & 2/3 & 1/2 & -1/2 \\ 0 & 1 & 0 & 0 \end{bmatrix} \begin{bmatrix} 1 & 0 & 0 & 0 \\ 0 & 1 & 0 & 0 \\ 0 & 0 & 1/2 & 0 \\ 0 & 0 & 0 & -1/2 \end{bmatrix} \begin{bmatrix} 1 & 0 & 0 & 0 \\ 0 & 0 & 0 & 1 \\ -1 & 1 & 1 & -1 \\ -1/3 & 1 & -1 & 1/3 \end{bmatrix}$$

which gives $\mathbf{P}^s$,

$$\begin{bmatrix} 1 & 0 & 0 & 0 \\ 2/3 & 1/3 & 1/2 & 1/2 \\ 1/3 & 2/3 & 1/2 & -1/2 \\ 0 & 1 & 0 & 0 \end{bmatrix} \begin{bmatrix} 1 & 0 & 0 & 0 \\ 0 & 1 & 0 & 0 \\ 0 & 0 & (1/2)^s & 0 \\ 0 & 0 & 0 & (-1/2)^s \end{bmatrix} \begin{bmatrix} 1 & 0 & 0 & 0 \\ 0 & 0 & 0 & 1 \\ -1 & 1 & 1 & -1 \\ -1/3 & 1 & -1 & 1/3 \end{bmatrix}$$

We finally get the limiting distribution by letting $s \to \infty$:

$$\begin{bmatrix} 1 & 0 & 0 & 0 \\ 2/3 & 1/3 & 1/2 & 1/2 \\ 1/3 & 2/3 & 1/2 & -1/2 \\ 0 & 1 & 0 & 0 \end{bmatrix} \begin{bmatrix} 1 & 0 & 0 & 0 \\ 0 & 1 & 0 & 0 \\ 0 & 0 & 0 & 0 \\ 0 & 0 & 0 & 0 \end{bmatrix} \begin{bmatrix} 1 & 0 & 0 & 0 \\ 0 & 0 & 0 & 1 \\ -1 & 1 & 1 & -1 \\ -1/3 & 1 & -1 & 1/3 \end{bmatrix} \quad (8.4.3)$$

$$= \begin{bmatrix} 1 & 0 & 0 & 0 \\ 2/3 & 0 & 0 & 1/3 \\ 1/3 & 0 & 0 & 2/3 \\ 0 & 0 & 0 & 1 \end{bmatrix}$$

8.5   TWO OTHER METHODS FOR FINDING LIMITING DISTRIBUTIONS

If the stochastic matrix $\mathbf{P}$ is simple enough, it is sometimes possible to calculate $\mathbf{P}^S$ directly, and thus find the limiting distribution. Consider the previous example,

$$\mathbf{P} = \begin{bmatrix} 1 & 0 & 0 & 0 \\ 1/2 & 0 & 1/2 & 0 \\ 0 & 1/2 & 0 & 1/2 \\ 0 & 0 & 0 & 1 \end{bmatrix}$$

Using the notation $p_{ij}(s) = P[X_{n+s} = j | X_n = i]$ we have the following:

$$p_{23}(s) = \tfrac{1}{2} p_{22}(s-1) = \tfrac{1}{2} \tfrac{1}{2} p_{23}(s-2)$$

Since $p_{23}(1) = 1/2$, $p_{23}(0) = 0$, it follows immediately that

$$p_{23}(s) = \begin{cases} (1/2)^s, & s \text{ odd} \\ 0, & s \text{ even} \end{cases} \qquad (8.5.1)$$

and

$$p_{22}(s) = \begin{cases} (1/2)^s, & s \text{ even} \\ 0, & s \text{ odd} \end{cases} \qquad (8.5.2)$$

so

$$\lim_{s \to \infty} p_{22}(s) = \lim_{s \to \infty} p_{23}(s) = 0$$

Since 1 is an absorbing state it is clear that

$$p_{21}(s) = \sum_{j=0}^{s-1} \frac{1}{2} p_{22}(j)$$

and we can conclude from this and (8.5.2),

$$\lim_{s \to \infty} p_{21}(s) = \frac{1}{2} \left(1 + \frac{1}{4} + \left(\frac{1}{4}\right)^2 + \cdots \right)$$

$$= \frac{1}{2} \left(\frac{1}{1 - 1/4} \right) = \frac{2}{3}$$

Similarly since 4 is an absorbing state

$$p_{24}(s) = \sum_{j=0}^{s-1} \frac{1}{2} p_{23}(j)$$

which by (8.5.1) gives

$$\lim_{s \to \infty} p_{24}(s) = 1/3$$

By symmetry, we find the remaining transition probabilities,

$$p_{22}(s) = p_{33}(s)$$
$$p_{23}(s) = p_{32}(s)$$
$$p_{21}(s) = p_{34}(s)$$
$$p_{24}(s) = p_{31}(s)$$

which shows that **P** has the limiting distribution (8.4.3).

Another way of looking at this problem is to consider the function

$$R(k) = P[X_s = 1 \text{ for some } s | X_0 = k]$$

which is the probability that the process, starting from state k, at some stage gets to state 1, where it stays from that time on. This function satisfies

$$R(1) = 1, \qquad R(4) = 0$$

and we have

$$R(2) = \frac{1}{2} R(1) + \frac{1}{2} R(3) = \frac{1}{2} + \frac{1}{2} R(3),$$

$$R(3) = \frac{1}{2} R(2) + \frac{1}{2} R(4) = \frac{1}{2}\left(\frac{1}{2} + \frac{1}{2} R(3)\right)$$

so that

$$R(2) = 2/3, \qquad R(3) = 1/3$$

By symmetry

$$Q(k) = P[X_s = 4 \text{ for some } s | X_0 = k]$$

has

$$Q(1) = 0, \qquad Q(2) = 1/3, \qquad Q(3) = 2/3, \qquad Q(4) = 1$$

which also verifies the limiting distribution (8.4.3).

8.6   RANDOM WALK

The term random walk stands for any Markov chain X_n, with the integers as its state space, whose transition probabilities satisfy

$$p_{ij} = \begin{cases} 1 - q_i, & j = i - 1 \\ q_i, & j = i + 1 \\ 0, & \text{else} \end{cases} \tag{8.6.1}$$

where $0 \le q_i \le 1$. Thus a random walk at each step must jump to one of the two neighboring states. One well-known example is when $q_i = \frac{1}{2}$, for all i. This can be physically realized by the net profit of a gambler who bets on each toss of a fair coin, winning a dollar if "heads" comes up, losing a dollar if "tails" comes up. In this case, X_n turns out to be a discrete approximation to the Wiener process, in the following sense.

Interpreting X_n as the net number of jumps of size ϵ of a particular after n time intervals of size τ, the question arises as to how to scale τ and ϵ so that one obtains a limit when $\epsilon \to 0$, $n \to \infty$. Since for the Wiener process,

$$E[(X(t) - X(0))^2] = 2Dt$$

a good choice would be

$$\epsilon^2 = 2D\tau.$$

Setting $\tau = t/n$, $\epsilon = \sqrt{2Dt/n}$, $X(t) = \epsilon X_n$, we obtain by the central limit theorem,

$$\lim_{n \to \infty} P[a < X(t) \le b] = \frac{1}{2\sqrt{\pi Dt}} \int_a^b e^{-x^2/4Dt} \, dx \tag{8.6.2}$$

Further, it can be verified that this limiting procedure yields a transition density satisfying the Fokker–Planck equation for the Wiener process.

Consider $p_{ij}(s)$ and note that

$$p_{ij}(s + 1) = \tfrac{1}{2}[p_{ij-1}(s) + p_{ij+1}(s)] \tag{8.6.3}$$

or

$$p_{ij}(s + 1) - p_{ij}(s) = \tfrac{1}{2}[p_{ij-1}(s) - 2p_{ij}(s) + p_{ij+1}(s)]$$

This, by defining

$$p(x, t|y) = p_{ij}(s)/\epsilon, \qquad t = s\tau, \qquad x = \epsilon j, \qquad y = \epsilon i$$

and dividing by τ, becomes

$$\frac{p(x, t + \tau|y) - p(x, t|y)}{\tau} = \frac{\epsilon^2}{2\tau} \frac{[p(x - \epsilon, t|y) - 2p(x, t|y) + p(x + \epsilon, t|y)]}{\epsilon^2}$$

$$\tag{8.6.4}$$

which in the limit as $n \to \infty$, $(\epsilon, \tau \to 0)$ is

$$\frac{\partial p}{\partial t} = D \frac{\partial^2 p}{\partial x^2} \tag{8.6.5}$$

8.7   RANDOM WALK WITH ABSORBING BARRIERS

As an illustration of the techniques that can be employed to handle stochastic matrices of arbitrary dimensions, we present the following generalization of one of the previous examples. Consider the Markov chain X_k modeling the "random walk with absorbing barriers" whose state space is $\{0, 1, \ldots, n + 1\}$ and

$$P[X_{k+1} = j | X_k = i] = p_{ij} = \begin{cases} \frac{1}{2}, & i = 1, \ldots, n, \quad j = i + 1 \text{ or } i - 1 \\ 1, & i = j = 0 \text{ or } i = j = n + 1 \\ 0, & \text{else} \end{cases} \tag{8.7.1}$$

The corresponding stochastic matrix is

$$\mathbf{P} = \begin{bmatrix} 1 & 0 & 0 & 0 & \cdots & 0 & 0 & 0 & 0 \\ 1/2 & 0 & 1/2 & 0 & \cdots & 0 & 0 & 0 & 0 \\ 0 & 1/2 & 0 & 1/2 & \cdots & 0 & 0 & 0 & 0 \\ \vdots & \vdots & \vdots & \vdots & \vdots & \vdots & \vdots & \vdots & \vdots \\ 0 & 0 & 0 & 0 & \cdots & 1/2 & 0 & 1/2 & 0 \\ 0 & 0 & 0 & 0 & \cdots & 0 & 1/2 & 0 & 1/2 \\ 0 & 0 & 0 & 0 & \cdots & 0 & 0 & 0 & 1 \end{bmatrix} \tag{8.7.2}$$

but we consider instead the matrix $\mathbf{P}'$ formed by deleting the first and last rows and columns from $\mathbf{P}$:

$$\mathbf{P}' = \begin{bmatrix} 0 & 1/2 & 0 & \cdots & 0 & 0 & 0 \\ 1/2 & 0 & 1/2 & \cdots & 0 & 0 & 0 \\ \vdots & \vdots & \vdots & \vdots & \vdots & \vdots & \vdots \\ 0 & 0 & 0 & \cdots & 1/2 & 0 & 1/2 \\ 0 & 0 & 0 & \cdots & 0 & 1/2 & 0 \end{bmatrix} \tag{8.7.3}$$

One can easily check that $(\mathbf{P}^s)' = (\mathbf{P}')^s$, so this matrix gives nearly the same information as $\mathbf{P}$. However, $\mathbf{P}'$ has the advantage of being symmetric, and therefore diagonalizable. The matrix $\mathbf{P}'$ is not stochastic, but satisfies

$$\sum_j p_{ij} \leq 1 \tag{8.7.4}$$

and

$$p_{ij} \geq 0 \qquad (8.7.5)$$

Such matrices are called substochastic matrices, and one can easily check that the same argument which we used for stochastic matrices implies that any eigenvalues of a substochastic matrix must have $|\lambda| \leq 1$. To find the diagonalization of $\mathbf{P}'$ we need to solve

$$\mathbf{P}'\mathbf{x} = \lambda\mathbf{x}$$

or

$$\begin{aligned}
\tfrac{1}{2}x_2 &= \lambda x_1 \\
\tfrac{1}{2}x_1 + \tfrac{1}{2}x_3 &= \lambda x_2 \\
\vdots \quad &= \quad \vdots \\
\tfrac{1}{2}x_{n-2} + \tfrac{1}{2}x_n &= \lambda x_{n-1} \\
\tfrac{1}{2}x_{n-1} &= \lambda x_n
\end{aligned}$$

This can be done by considering instead the corresponding infinite system

$$\tfrac{1}{2}x_k + \tfrac{1}{2}x_{k-2} = \lambda x_{k-1}, \qquad k = \ldots, -1, 0, 1, \ldots \qquad (8.7.6)$$

subject to the "boundary" conditions

$$x_0 = x_{n+1} = 0 \qquad (8.7.7)$$

This system has particular solutions of the form

$$x_k = \rho^k$$

where

$$\rho = \lambda \pm \sqrt{\lambda^2 - 1}$$

or, noting $|\lambda| \leq 1$, and setting $\lambda = \cos\theta$,

$$= \cos\theta \pm i\sin\theta = e^{\pm i\theta}$$

The general solution is therefore

$$x_k = Ae^{i\theta k} + Be^{-i\theta k} \qquad (8.7.8)$$

If we impose the boundary conditions to this equation, we get

$$A + B = 0 \qquad\qquad (8.7.9)$$
$$A^{i\theta(n+1)} + Be^{-i\theta(n+1)} = 0$$

or

$$\sin(\theta(n + 1)) = 0$$

which determines

$$\theta = \frac{m\pi}{n + 1}, \qquad m = 1, 2, \ldots, n$$

From this we conclude the eigenvalues of $\mathbf{P}'$ are

$$\lambda_m = \cos\frac{m\pi}{n + 1}$$

and for each m, the corresponding eigenvectors are given by

$$x_k(m) = C \sin\frac{km\pi}{n + 1}$$

where C is any constant. To obtain orthonormal eigenvectors, and thus the orthogonal matrix that diagonalizes $\mathbf{P}'$, we solve

$$1 = \sum_{k=1}^{n} x_k^2(m) = C^2 \sum_{k=1}^{n} \sin^2\frac{km\pi}{n + 1}$$

which by the identity

$$\sum_{k=1}^{n} \sin^2\frac{km\pi}{n + 1} = \frac{n + 1}{2}$$

gives $C = \sqrt{2/(n + 1)}$.

To show this identity, first note that

$$\sum_{k=1}^{n} \sin^2\frac{km\pi}{n + 1} = \frac{1}{2} \sum_{k=0}^{n} \left(1 - \cos\frac{2km\pi}{n + 1}\right)$$

$$= \frac{n + 1}{2} - \frac{1}{2} \sum_{k=0}^{n} \cos\frac{2km\pi}{n + 1}$$

Since

$$\exp\left(\frac{2m\pi}{n+1}\right) \sum_{k=0}^{n} \exp\left(\frac{2mk\pi}{n+1}\right) = \sum_{k=0}^{n} \exp\left(\frac{2mk\pi}{n+1}\right)$$

implies

$$\sum_{k=0}^{n} \exp\left(\frac{2mk\pi}{n+1}\right) = 0$$

we have

$$\sum_{k=0}^{n} \cos\frac{2km\pi}{n+1} = \text{Re}\left[\sum_{k=0}^{n} \exp\left(\frac{2mk\pi}{n+1}\right)\right] = \text{Re}[0] = 0$$

Now that we know the eigenvalues and an orthonormal set of eigenvectors for $\mathbf{P}'$, we can write its diagonalization

$$\mathbf{P}' = \mathbf{R}^T\mathbf{D}\mathbf{R}$$

where $\mathbf{D}$ is diagonal with diagonal entries

$$\lambda_1 = \cos\frac{\pi}{n+1}, \ldots, \lambda_n = \cos\frac{n\pi}{n+1} \tag{8.7.10}$$

and $\mathbf{R}$ is the matrix whose mth row is the eigenvector

$$x_k(m) = \sqrt{\frac{2}{n+1}} \sin\frac{km\pi}{n+1} \tag{8.7.11}$$

Therefore, $(\mathbf{P}')^s = \mathbf{R}^T\mathbf{D}^s\mathbf{R}$ has entries

$$p_{ij}'(s) = \sum_{k=1}^{n} \lambda_k^s x_i(k)x_j(k) \tag{8.7.12}$$

and since $|\lambda_k| < 1$, $k = 1, \ldots, n$, we conclude that

$$\lim_{s \to \infty} (\mathbf{P}')^s = \mathbf{0} \tag{8.7.13}$$

To find the remaining entries of $\mathbf{P}^s$, note that the first and last rows of $\mathbf{P}^s$ are the same as those of $\mathbf{P}$, so only the first and last columns remain to be specified. These entries can be found by the methods of the first part of Section 8.5. Finding the limiting distribution is a little easier. We consider, as before,

$$R(k) = P[X_s = 0 \text{ for some } s | X_0 = k] \tag{8.7.14}$$

and obtain the recursion relation

$$R(k) = \tfrac{1}{2}R(k + 1) + \tfrac{1}{2}R(k - 1), \qquad k = 1, 2, \ldots, n \tag{8.7.15}$$

with the boundary conditions

$$R(0) = 1, \qquad R(n + 1) = 0$$

This has the unique solution

$$R(k) = 1 - \frac{k}{n + 1} \tag{8.7.16}$$

and thus

$$\lim_{s \to \infty} \mathbf{P}^s = \begin{bmatrix} 1 & 0 & 0 & \cdots & 0 & 0 & 0 \\ n/(n+1) & 0 & 0 & \cdots & 0 & 0 & 1/(n + 1) \\ \vdots & \vdots & \vdots & \vdots & \vdots & \vdots & \vdots \\ 2/(n + 1) & 0 & 0 & \cdots & 0 & 0 & (n - 1)/(n + 1) \\ 1/(n + 1) & 0 & 0 & \cdots & 0 & 0 & n/(n + 1) \\ 0 & 0 & 0 & \cdots & 0 & 0 & 1 \end{bmatrix} \tag{8.7.17}$$

It is interesting to look at the diagonalization of $\mathbf{P}'$ from a different perspective. For a given initial distribution $\mathbf{y} = (y_0, \ldots, y_{n+1})$, define $\mathbf{y}' = (y_1, \ldots, y_n)$, and consider

$$\mathbf{y}'^{\mathrm{T}}\mathbf{P}' = \mathbf{y}'^{\mathrm{T}}\mathbf{R}^{\mathrm{T}}\mathbf{DR}$$

Interpreting $\mathbf{R}$ as a change of coordinates into a basis of eigenvectors, this equation says that you can find the distribution on states $1, \ldots, n$ by writing $\mathbf{y}'$ as a linear combination of (left) eigenvectors $\{\mathbf{x}(i), i = 1, \ldots, n\}$

$$\mathbf{y}' = \sum a_m \mathbf{x}(m)$$

so that

$$\mathbf{y}'^{\mathrm{T}}\mathbf{P}' = \sum a_m \lambda_m \mathbf{x}^{\mathrm{T}}(m)$$

Here the a_m are the entries in the matrix product $\mathbf{y}'^T \mathbf{R}^T$. Further, we can conclude that

$$\mathbf{y}'^T (\mathbf{P}')^s = \sum a_m (\lambda_m)^s \mathbf{x}^T(m)$$

We can imitate this to help us solve the related Fokker–Planck equation of the "Brownian motion with absorbing boundaries",

$$\frac{\partial P}{\partial t} = D \frac{\partial^2 P}{\partial x^2} \qquad (8.7.18)$$

with the boundary conditions

$$P(x, t) = 0, \; x \notin (0, a) \qquad (8.7.19)$$

and initial condition

$$P(x, 0) = P_0 \qquad (8.7.20)$$

This is the heat equation in one dimension, modeling the temperature of an infinite uniform slab with a heat sink on each side.

By analogy, we consider the "eigenfunctions"

$$f_m(x) = \sin \left(\frac{2\pi m x}{a} \right) \qquad (8.7.21)$$

and replacing λ^s by its continuous time analogue $e^{-\lambda t}$, we expect

$$g_m(x) = e^{-\lambda t} \sin \left(\frac{2\pi m x}{a} \right)$$

to be a particular solution of (8.7.18) for an appropriate choice of λ, depending on m. Substituting this into the differential equation yields

$$\lambda = D \left(\frac{2\pi m}{a} \right)^2$$

Therefore by writing the initial condition in a Fourier sine series

$$P_0(x) = \sum_{m=1}^{\infty} b_m \sin \left(\frac{2\pi m x}{a} \right)$$

we can find the density at time t, by the formula

$$P(x, t) = \sum_{m=1}^{\infty} \exp\left(-Dt\left(\frac{2\pi m}{a}\right)^2\right) b_m \sin\left(\frac{2\pi mx}{a}\right) \qquad (8.7.22)$$

8.8   THE STATIONARY DISTRIBUTION AND MEAN RECURRENCE TIME

Under general conditions there is a unique stationary distribution for a Markov chain, given in terms of the mean recurrence time $\mu(j)$, which is the expected length of time it takes the process, starting from a state S_j, to return to that state. In this section we investigate this important property of Markov chains.

Classification of states

Suppose that X_n is a Markov chain. A state S_i is said to be **transient** if the probability of eventually returning to that state is less than 1:

$$P[X_n = S_i \text{ for some } n > 0 | X_0 = S_i] < 1$$

and **recurrent** if the probability of return is equal to 1:

$$P[X_n = S_i \text{ for some } n > 0 | X_0 = S_i] = 1$$

We define the **first hitting time** T_i to a state S_i by

$$T_i = \min \{n: n > 0, X_n = S_i\} \qquad (8.8.1)$$

and the **mean recurrence time** $\mu(i)$ of a state i by

$$\mu(i) = \begin{cases} \infty, & \text{if } i \text{ is transient} \\ \sum_{k=1}^{\infty} kP[T_i = k | X_0 = S_i], & \text{if } i \text{ is recurrent} \end{cases}$$

Thus the mean recurrence time is the average number of steps for the Markov chain, starting from i, to return to the state i. A recurrent state is called **positive recurrent** if its mean recurrence time is finite, and **null recurrent** if its mean recurrence time is infinite. Every state is either transient, positive recurrent, or null recurrent. A state S_i is said to **communicate** with a state S_j if there exists $k > 0$ such that

$$p_{ij}(k) > 0$$

Two states are said to **intercommunicate** if each communicates with the other. If S_i and S_j intercommunicate, they are of the same type.

Sets of states

A set of states $E = \{S_j\}$ is called **irreducible** if any two states in E intercommunicate, it is called **closed** if no states in E communicate with any states not in E. It follows from what we have just seen that irreducible sets of states are all of the same type. Markov chains whose state spaces have a certain property are said to have that property. For example, if the state space is irreducible, the chain is called irreducible.

The irreducible recurrent set F of all states intercommunicating with a given recurrent state is closed. To show this, suppose that S_j is a state in F which communicates with a state S_k not in F. Then there is positive probability that starting at S_j the chain reaches S_k before it returns to S_j. Once it reaches S_k it cannot return to S_j. This contradicts recurrence.

It follows that the state of space of a Markov chain can be decomposed into disjoint sets, one being the collection of all transient states, and the others irreducible closed sets of recurrent states.

Finite-state spaces

Suppose a Markov chain has $N < \infty$ states and is irreducible. In this case we can show that all the states are recurrent, since then for each i there exists $k \geq 0$ with $p_{ii}(k) > 0$, so

$$P[X_n \neq S_i \text{ for all } n < mk | X_0 = S_i] < (1 - p_{ii}(k))^m \to 0 \qquad \text{as } m \to \infty \quad (8.8.2)$$

By using this one can also show that the mean recurrence time is finite.

With these assumptions there exists a unique stationary distribution $\pi(i)$ which can be described in terms of the mean recurrence times by

$$\pi(i) = \frac{1}{\mu(i)} \quad (8.8.3)$$

Since the stationary distribution is usually easy to find, this formula gives an easy way to find the mean recurrence times. We first show uniqueness.

Suppose that $\pi = (\pi(1), \ldots, \pi(N))$ is a stationary distribution for the Markov process, so that

$$\sum_{i=1}^{N} \pi(i)p_{ik} = \pi(k)$$

By setting $P[X_0 = S_i] = \pi(i)$ we can verify the following

$$
\begin{aligned}
\mu(j)\pi(j) &= \sum_{k=1}^{\infty} kP[T_j = k | X_0 = S_j]P[X_0 = S_j] \\
&= \sum_{k=1}^{\infty} P[T_j \geq k | X_0 = S_j]P[X_0 = S_j] \\
&= \sum_{k=1}^{\infty} P[T_j \geq k, X_0 = S_j] \\
&= P[X_0 = S_j] + \sum_{k=2}^{\infty} P[X_0 = S_j, X_1 \neq S_j, \ldots, X_{k-1} \neq S_j] \\
&= P[X_0 = X_j] + \sum_{k=2}^{\infty} (P[X_1 \neq S_j, \ldots, X_{k-1} \neq S_j] \\
&\quad - P[X_0 \neq S_j, X_1 \neq S_j, \ldots, X_{k-1} \neq S_j])
\end{aligned}
$$

by stationarity of the initial distribution

$$
\begin{aligned}
&= P[X_0 = S_j] + \sum_{k=2}^{\infty} (P[X_0 \neq S_j, \ldots, X_{k-2} \neq S_j] \\
&\quad - P[X_0 \neq S_j, X_1 \neq S_j, \ldots, X_{k-1} \neq S_j])
\end{aligned}
$$

the sum telescopes

$$
\begin{aligned}
&= P[X_0 = S_j] + P[X_0 \neq S_j] \\
&\quad - \lim_{k \to \infty} P[X_0 \neq S_j, X_1 \neq S_j, \ldots, X_k \neq S_j]
\end{aligned}
$$

which by recurrence

$$
\begin{aligned}
&= P[X_0 = S_j] + P[X_0 \neq S_j] \\
&= 1
\end{aligned}
$$

This shows that the only candidate for the stationary distribution is (8.8.3).

To show existence, define $a_i(m)$ to be the expected number of visits of the chain to state i between visits to state m

$$
a_i(m) = \sum_{k=0}^{\infty} P[X_k = S_i, T_m > k | X_0 = S_m]
$$

so that one would expect

$$\mu(i)a_i(m) = \mu(m)$$

or

$$\frac{1}{\mu(i)} = \frac{a_i(m)}{\mu(m)} \tag{8.8.4}$$

We can show existence, and obtain (8.8.4) as a by-product, by verifying

$$a_j(m) = \sum_{i=1}^{N} a_i(m)p_{ij} \tag{8.8.5}$$

and

$$\sum_{i=1}^{N} a_i(m) = \mu(m) \tag{8.8.6}$$

The second statement is easy:

$$\mu(m) = \sum_{k=0}^{\infty} P[T_m > k | X_0 = S_m]$$

$$= \sum_{i=1}^{N} \sum_{k=0}^{\infty} P[X_k = S_i, T_m > k | X_0 = S_m]$$

$$= \sum_{i=1}^{N} a_i(m)$$

The first statement follows if $j \neq m$ by writing

$$a_j(m) = \sum_{k=0}^{\infty} P[X_k = S_j, T_m > k | X_0 = S_m]$$

$$= \sum_{k=1}^{\infty} P[X_k = S_j, T_m > k | X_0 = S_m]$$

since $j \neq m$

$$= \sum_{k=1}^{\infty} P[X_k = S_j, T_m > k - 1 | X_0 = S_m]$$

$$= \sum_{k=1}^{\infty} \sum_{i=1}^{N} P[X_k = S_j, X_{k-1} = S_i, T_m > k - 1 | X_0 = S_m]$$

by the Markov property

$$= \sum_{k=1}^{\infty} \sum_{i=1}^{N} p_{ij} P[X_{k-1} = S_i, T_m > k - 1 | X_0 = S_m]$$

$$= \sum_{i=1}^{N} p_{ij} \sum_{k=0}^{\infty} P[X_k = S_i, T_m > k | X_0 = S_m]$$

$$= \sum_{i=1}^{N} p_{ij} a_i(m)$$

If $j = m$ we write

$$a_m(m) = 1$$

which by recurrence

$$= \sum_{k=1}^{\infty} P[T_m = k | X_0 = S_m]$$

$$= \sum_{k=1}^{\infty} \sum_{i=1}^{N} P[X_{k-1} = S_i, T_m = k | X_0 = S_m]$$

$$= \sum_{k=1}^{\infty} \sum_{i=1}^{N} p_{ik} P[T_m > k - 1 | X_0 = S_m]$$

$$= \sum_{i=1}^{N} p_{ik} a_i(m)$$

We apply this result in the next section.

General state spaces

The importance of positive recurrence is contained in the result that a general irreducible Markov chain has a stationary distribution if and only if it is positive recurrent. Suppose that $\pi(k)$ is a stationary distribution of a Markov chain, so that

$$\sum_{k=1}^{\infty} \pi(k) p_{kj}(n) = \pi(j)$$

Setting

$$g_{ij}(m) = \sum_{n=1}^{m} p_{ij}(n)$$

the preceding equation gives

$$\sum_{k=1}^{\infty} \pi(k) \frac{g_{kj}(m)}{m} = \pi(j)$$

We shall see the end of Section (8.12) that if S_j is transient or null recurrent, $g_{kj}(m)/m \rightarrow 0$ as $m \rightarrow \infty$. This implies that $\pi(k) = 0$ for all k, so that there is no stationary distribution. If the chain is positive recurrent, the unique stationary distribution is given by

$$\pi_i = \frac{1}{\mu(i)}$$

We omit the details of the proof, which is similar to the finite-state-space case.

8.9 THE EHRENFEST MODEL REVISITED, ELASTIC RANDOM WALK

In this section we discuss a random walk, which in the limit yields the Ornstein–Uhlenbeck process. We consider a random walk X_n on $-R, \ldots, R$ with

$$P[X_{n+1} = j | X_n = i] = \begin{cases} \frac{1}{2}(1 - i/R), & j = i + 1 \\ \frac{1}{2}(1 + i/R), & j = i - 1, \quad i, j \in \{-R, \ldots, R\} \\ 0, & \text{else} \end{cases}$$

(8.9.1)

The random variable $X_n + R$ is the one considered in the Ehrenfest model. We have the relation

$$p_{ij}(s + 1) = \frac{R + j + 1}{2R} p_{ij+1}(s) + \frac{R - j + 1}{2R} p_{ij-1}(s) \qquad (8.9.2)$$

which under appropriate assumptions formally becomes

$$\frac{\partial p}{\partial x} = \gamma \frac{\partial (xp)}{\partial x} + D \frac{\partial^2 p}{\partial x^2} \qquad (8.9.3)$$

To show this write (8.9.2) as

$$p_{ij}(s + 1) - p_{ij}(s) = \frac{1}{2} [p_{ij+1}(s) - 2p_{ij}(s) + p_{ij-1}(s)]$$

$$+ \frac{j + 1}{2R} p_{ij+1}(s) - \frac{j - 1}{2R} p_{ij-1}(s)$$

(8.9.4)

Using, as before, $\epsilon^2 = 2D\tau$, $\tau = t/s$, $\epsilon = \sqrt{2Dt/s}$ and the notation

$$p(x, t|y) = p_{ij}(s)/\epsilon, \qquad t = s\tau, \qquad x = \epsilon j, \qquad y = \epsilon i \qquad (8.9.5)$$

this is

$$\frac{p(x, t + \tau|y) - p(x, t|y)}{\tau} = \frac{\epsilon^2}{2\tau} \frac{p(x + \epsilon, t|y) - 2p(x, t|y) + p(x - \epsilon, t|y)}{\epsilon^2}$$

$$+ \frac{1}{R\tau} \frac{(x + \epsilon)p(x + \epsilon, t|y) - (x - \epsilon)p(x - \epsilon, t|y)}{2\epsilon}$$

$$(8.9.6)$$

which by assuming $\gamma = 1/R\tau$ approaches a fixed number as $s \to \infty$ becomes

$$\frac{\partial p}{\partial x} = \gamma \frac{\partial(xp)}{\partial x} + D \frac{\partial^2 p}{\partial x^2} \qquad (8.9.7)$$

The Ehrenfest model was proposed as a model of heat transfer, and we are now in a position to reconcile the irreversibility of the laws of thermodynamics and the reversibility of the Ehrenfest chain. Suppose that we start off with only 100 balls in one box and none in the other, and perform 1 million selections every second. What is the expected waiting time until we see that configuration again? By the preceding section, since the stationary distribution (8.1.3) gives us

$$\pi(100) = (\tfrac{1}{2})^{100} \qquad (8.9.0)$$

by (8.8.3) the mean recurrence time is therefore 2^{100} millionth of a second, which is 4×10^{16} years! The interpretation of this statement is that although large quantities of "heat" can flow in the "wrong" direction we must wait a very long time to see it happen. The mean recurrence time for states closer to equilibrium on the other hand are small enough that we can expect to witness the recurrence.

8.10   BRANCHING PROCESSES

As an example of the use of the probability generating function in Markov chains, we consider the branching process. This is sometimes taken as a model for population growth, and has been used to examine the question of extinction of family names.

Let X_n be the number of individuals in the nth generation of a population. For the purposes of the example, consider only the male individuals in the population. Assume that all the individuals act independently, and have the same probability distribution for the number of offspring. This means that the individual family sizes are independent identically distributed random variables Y, which can take on the values $0, 1, 2, \ldots$, and

$$X_{n+1} = \sum_{k=1}^{X_n} Y_k$$

where the Y_k's are independent random variables with the same distribution as Y. The sequence X_n is then a Markov chain.

To analyze this process, since X_n is a sum of independent random variables, an approach using either the characteristic function or the moment generating function would seem a good choice. However, it turns out to be more convenient to use a variant of the moment generating function, called the probability generating function. This is defined for a random variable X by

$$G_X(t) = E[t^X]$$

The reason the probability generating function is useful stems from the following result about random sums of random variables. Suppose that Y is a random variable, and Y_k are independent random variables with the same distribution as Y. If N is a positive integer valued random variable, independent of the Y_k's, setting

$$S = \sum_{k=1}^{N} Y_k$$

then

$$G_S(t) = G_N(G_Y(t))$$

This is easy to show:

$$G_S(t) = E[t^S]$$

$$= E[t^{\sum_{k=1}^N Y_k}]$$

$$= \sum_{n=1}^{\infty} E[t^{\sum_{k=1}^n Y_k}]P[N = n]$$

$$= \sum_{n=1}^{\infty} (E[t^Y])^n P[N = n]$$

$$= G_N(E[t^Y]) = G_N(G_Y(t))$$

Applying this result to the branching process, we get

$$G_{X_{n+1}}(t) = G_{X_n}(G_Y(t))$$

Assuming that we start with one individual, $X_0 = 1$, $X_1 = Y$, this means that

$$G_{X_n}(t) = G_Y(G_Y(\cdots (G_Y(t)) \cdots))$$

where there are n occurrences of G_Y in the convolution. Therefore, we need only study G_Y and its iterates.

The properties of $G_Y(t)$ are easy to find once you note that

$$G_Y(t) = E[t^Y]$$

$$= \sum_{k=0}^{\infty} P[Y = k]t^k$$

is a power series that can be differentiated term by term for $t \in [0, 1]$. This yields that on $[0, 1]$ $G_Y(t)$ is a continuous nondecreasing nonnegative function, which is convex, since

$$G_Y''(t) = \sum_{k=2}^{\infty} k(k - 1)P[Y = k]t^{k-2}$$

is nonnegative. Further, $G_Y(0) = P[Y = 0]$ and $G_Y(1) = 1$.

We are now in a position to answer the question of extinction of family names. Let

$$e_n = P[X_n = 0]$$

be the probability of "extinction" at generation n. Notice that e_n is a nondecreasing sequence, so it has a limit e as $n \to \infty$ with $0 \le e \le 1$. Now

$$e_k = G_{X_k}(0) = G_Y(G_{X_{k-1}}(0)) = G_Y(e_{k-1})$$

so by the continuity of G_Y, e is a root of

$$t = G_Y(t)$$

with $0 \le t \le 1$. By the properties of G_Y, any such t has

$$G_Y(0) \le G_Y(t)$$

and

$$G_Y(\cdots (G_Y(0)) \cdots) \le G_Y(\cdots (G_Y(t)) \cdots)$$

or

$$e_k \le t$$

so

$$e \leq t$$

Thus e is the minimal root of

$$t = G_Y(t)$$

in $[0, 1]$.

If we consider the graph of $G_Y(t) = y$, it is clear that there are three cases to consider (see Fig. 8.1):

(i)   If $P[Y = 0] = 0$, it is clear that extinction cannot occur, and in this case 0 is the minimal solution of $G_Y(t) = t$.

(ii)   If $P[Y = 0] > 0$ and $G'(1) > 1$, there will be a solution to $G_Y(t) = t$ with $0 < t < 1$, and correspondingly the probability of extinction will be positive, but not equal to 1.

(iii)   If $G'(1) \leq 1$, then the probability of extinction is 1 unless $G_Y(t) = t$, which means $P[Y = 0] = 0$ and (i) implies that extinction cannot occur.

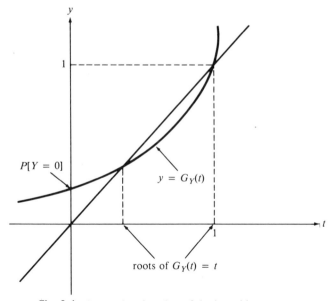

Fig. 8.1. Generating function of the branching process

Since

$$G'(1) = E[Y]$$

is the average family size, one can therefore predict whether the probability of extinction is 1, zero, or between 1 and zero, based on the probability of childlessness, and the average family size. The initially surprising fact that the chances of extinction are at least as large as the chances of childlessness, is explained by the assumption that $X_0 = 1$. If the initial number of individuals in the "clan" is greater than 1, $X_0 = m > 1$, by independence, the probability of extinction is e^m.

The calculation of the distribution of X_n for a given Y is difficult. The moments are easy to find, however. This follows from

$$\left. \frac{d^n}{dt^n} G_X(t) \right|_{t=1} = E[X(X - 1) \cdots (X - n + 1)]$$

Since the derivatives of G_Y and the derivatives of G_{X_n} are related, as we can see from

$$\left. \frac{d}{dt} G_{X_2}(t) \right|_{t=1} = \left. \frac{d}{dt} G_Y(G_Y(t)) \right|_{t=1} = \left. G_Y'(G_Y(t))G_Y'(t) \right|_{t=1} = (G_Y'(1))^2$$

the moments of X_n can be found in terms of the moments of Y.

8.11   CHARACTERIZING TRANSIENT AND RECURRENT STATES

The behavior of general Markov chains with stationary transition probabilities can be more complicated than the examples we have seen so far. In general the matrix of transition probabilities is not diagonalizable, so the techniques we have developed at the beginning of the chapter do not apply. For the remainder of this chapter, we discuss some of the properties of Markov chains which can be deduced without using diagonalization.

Recall that we defined the hitting time to a state S_i by

$$T_i = \min \{n: n > 0, X_n = S_i\}$$

and said that a state S_i is called transient if the probability of eventual return to that state is less than 1,

$$P[T_i < \infty | X_0 = S_i] < 1$$

and recurrent if the probability of eventual return is equal to 1,

$$P[T_i < \infty | X_0 = S_i] = 1$$

If we define the **hitting probabilities** f_{ij} of the chain by

$$f_{ij} = P[T_j < \infty | X_0 = S_i]$$

then S_i is transient if

$$f_{ii} < 1$$

and recurrent if

$$f_{ii} = 1$$

We use the letter f here because these are sometimes referred to as the **first passage probabilities.**

The transience or recurrence of a state can be characterized in terms of the **potential matrix G** defined by

$$\mathbf{G} = \sum_{k=1}^{\infty} \mathbf{P}^k$$

whose entries are

$$g_{ij} = \sum_{k=1}^{\infty} p_{ij}(k)$$

The entries g_{ij} of **G** are the average number of times the process visits the state S_j given it starts in state S_i. By defining N_j to be the random variable that counts the total number of visits the Markov chain makes to state S_j

$$N_j = \sum_{k=1}^{\infty} 1_{S_j}(X_k)$$

we have

$$g_{ij} = E[N_j | X_0 = S_i]$$

Further, since the process starts afresh upon arriving at S_j,

$$P[N_j \geq 2 | X_0 = S_i] = f_{ij} f_{jj}$$

and similarly

$$P[N_j \geq k | X_0 = S_i] = f_{ij}(f_{jj})^{k-1}$$

Suppose S_j is recurrent, so $f_{jj} = 1$, then

$$P[N_j = \infty | X_0 = S_i] = f_{ij}$$

If $f_{ij} > 0$, this means

$$P[N_j = \infty | X_0 = S_i] > 0$$

which implies

$$E[N_j | X_0 = S_i] = g_{ij} = \infty$$

If $f_{ij} = 0$, then of course

$$g_{ij} = 0$$

On the oher hand if S_j is transient, so that $f_{jj} < 1$, then

$$P[N_j \geq k | X_0 = S_i] = f_{ij}(f_{jj})^{k-1}$$

implies

$$g_{ij} = E[N_j | X_0 = S_i] = \sum_{k=1}^{\infty} k f_{ij}(f_{jj}^{k-1} - f_{jj}^k) = f_{ij}(1 + f_{jj} + f_{jj}^2 + f_{jj}^3 + \cdots)$$

which is

$$g_{ij} = \frac{f_{ij}}{1 - f_{jj}} < \infty$$

Thus a recurrent state is one which if it is visited at all is visited infinitely often. A transient state can only be visited a finite number of times. In particular, if S_j is a transient state, as $k \to \infty$, $p_{ij}(k) \to 0$, since

$$\sum_{k=1}^{\infty} p_{ij}(k) = g_{ij} < \infty$$

8.12   ERGODICITY

A process is called ergodic if it is possible to find some statistical property of the process by watching only a single outcome (one sample path). The most basic ergodic theorem is the **strong law of large numbers,** which says that if $X_1, X_2, \ldots$ are independent identically distributed random variables, then there exists μ such that almost surely

$$\lim_{n \to \infty} \frac{\sum_{k=1}^{n} X_k}{n} = \mu$$

if and only if $E[|X_1|] < \infty$, and $\mu = E[X_1]$. For coin tossing, this says, in the long run, that any sequence of tosses must average out to an equal proportion of heads and tails. This is the so-called "law of averages." It does not, however, tell how long you must wait for this equalization to occur. For physical systems, we may get to see only one realization of a supposed stochastic process. Any conclusions about the nature of the process must then be based on the assumption of ergodicity.

Suppose that X_n is a Markov chain. We say that X_n satisfies the **mean ergodic theorem** if it has a unique stationary distribution $\pi(j)$ and almost surely

$$\lim_{n \to \infty} \frac{\sum_{k=1}^{n} 1_{S_j}(X_k)}{n} = \pi(j)$$

This says that the average number of visits that the chain makes to each state approaches the stationary distribution of the chain.

Suppose that X_n is recurrent and irreducible. Introduce

$$N_j(m) = \sum_{k=1}^{m} 1_{S_j}(X_k)$$

and

$$g_{ij}(m) = \sum_{k=1}^{m} p_{ij}(k)$$

which count the number of visits in the first m steps, and the expected number of visits in m steps, respectively, given that the chain starts at S_i. Further, consider the time, $T_j(k)$, it takes for the process to visit the state S_j a total of k times

$$T_j(k) = \min \{m: m \geq 1, N_j(m) = k\}$$

By step n the process has made $N_j(n)$ visits to S_j so

$$T_j(N_j(n)) \leq n < T_j(N_j(n) + 1)$$

and

$$\frac{T_j(N_j(n))}{N_j(n)} \leq \frac{n}{N_j(n)} \leq \frac{T_j(N_j(n) + 1)}{N_j(n)}$$

By recurrence as $n \rightarrow \infty$ we know that $N_j(n) \rightarrow \infty$ almost surely. Further, by the strong law of large numbers, viewing the arrival at state S_j as a renewal process, almost surely

$$\lim_{k \rightarrow \infty} \frac{T_j(k)}{k} = \mu(j)$$

This is true even if the mean recurrence time is infinite. Thus almost surely

$$\lim_{n \rightarrow \infty} \frac{N_j(n)}{n} = \frac{1}{\mu(j)}$$

where $\mu(j)$ is the mean recurrence time for S_j. This says that irreducible positive recurrent chains satisfy the mean ergodic theorem. Taking expectations, we get

$$\lim_{n \rightarrow \infty} \frac{g_{ij}(n)}{n} = \frac{1}{\mu(j)}$$

so if X_n is null recurrent

$$\lim_{n \rightarrow \infty} \frac{g_{ij}(n)}{n} = 0$$

as needed for Section 8.8.

8.13 CONVERGENCE TO THE EQUILIBRIUM DISTRIBUTION

We have seen in our examples that in some situations $\mathbf{P}^k$ converges, in which case the rows of the limit are stationary distributions for the associated Markov chain. If the chain is irreducible, these must be the unique stationary distribution. In this section we show that we have this convergence for a specific class of chains.

Periodicity

A state S_j is said to be **periodic** with period N if

$$N = gcd\{n: p_{jj}(n) > 0\}$$

In an irreducible set of states all the states have the same period. The chain X_n cannot return to state S_j except when $n = Nk$, where k is a positive integer. If $N = 1$, the state is called **aperiodic.** The Ehrenfest model provides a ready example of a chain with periodic states. Periodicity makes it impossible for $\mathbf{P}^k$ to converge, although $\mathbf{P}^{Nk}$ can converge.

Aperiodic chains can be characterized by the property that for each k there exists an m such that

$$p_{kk}(n) > 0$$

for $n > m$. To show this, consider the set

$$D = \{n | p_{kk}(n) > 0\}$$

which is closed under addition. Since the chain is aperiodic the gcd of this set is 1. Consider the set L of all possible finite linear combinations of elements of D

$$\sum_{k=1}^{n} a_k d_k$$

where the a_k's are integers and n is arbitrary. Let d be its smallest positive element. Since L is closed under differences, it must be the set of all multiples of a number d

$$L = \{ad, a \in \mathbb{Z}\}$$

Thus d divides every d_j, so $d = 1$.

Now consider the sum from L which gives 1. Its positive entries and its negative entries differ by 1. Therefore D contains two consecutive elements, say x and $x + 1$, one of which is the sum of the positive entries, the other of which is absolute value of the sum of the negative entries. Since any integer larger than x^2 can be expressed as

$$ax + b(x + 1)$$

where a and b are positive integers, D contains any integer larger than x^2. Setting $m = x^2$ completes the result.

It follows immediately from this and the Chapmann–Kolmogorov equation, that if an aperiodic chain is irreducible, then for each pair of states S_i and S_j, there exists an m so that

$$p_{ij}(n) > 0$$

for any $n > m$.

Convergence

Suppose that a Markov chain is irreducible positive recurrent and aperiodic; we can show that the entries of $\mathbf{P}^n$ converge to the stationary distribution

$$p_{ij}(n) \to \pi(j)$$

as $n \to \infty$.

To do this we first define a Markov chain $\mathbf{X}_n = (X_n, Y_n)$, where X_n and Y_n are two independent copies of the original chain. Thus $\mathbf{X}_n$ has state space $\{S_{ij}\}$ consisting of ordered pairs

$$S_{ij} = (S_i, S_j)$$

of states from the original chain. The transition probabilities for this new chain are given by independence as

$$P[\mathbf{X}_{n+m} = (S_i, S_r) | \mathbf{X}_n = (S_j, S_q)] = p_{ij}(m)p_{rq}(m)$$

Let X_n have an arbitrary initial distribution, and Y_n the stationary initial distribution π.

Since the original chain is aperiodic, $\mathbf{X}_n$ is irreducible. It is also positive recurrent since it has a stationary distribution given by

$$P[\mathbf{X}_n = (S_i, S_j)] = \pi(i)\pi(j)$$

This implies

$$P[\mathbf{X}_n = (S_k, S_k) \text{ for some } n] = 1$$

almost surely, which says that

$$P[X_n = Y_n \text{ for some } n] = 1$$

almost surely. This implies that the distribution of X_n converges to the distribution of Y_n, which is the stationary distribution for all n, and we are done. Of course, the stationary distribution is given in terms of the mean recurrence times.

PROBLEMS

8.1. Suppose that X_n is the Markov chain determined by the Ehrenfest model.

(a) Show that the binomial distribution (8.1.3) is the unique stationary distribution.

(b) Show that in the limit described in section 8.9, this distribution becomes the stationary distribution for the associated Ornstein–Uhlenbeck process.

(c) Complete the verification that the Ehrenfest model is time reversible.

8.2. Consider the Ehrenfest model with $R = 2$.

(a) Find its matrix of transition probabilities $\mathbf{P}$.

(b) Compute $\mathbf{P}^2$, $\mathbf{P}^3$.

(c) What is the limit of $\mathbf{P}^n$ as $n \to \infty$?

8.3. Suppose that we have two boxes and $2R$ balls, R of which are white and R of which are black, and distribute them so that there are R balls in each box. A ball is selected at random from each box, and each ball is put on the opposite box. Suppose X_n denotes the number of white balls in the first box after n repetitions.

(a) Find the transition matrix for this Markov chain.

(b) Find its stationary distribution.

(c) Is this Markov chain time reversible?

8.4. What form does the Chapmann–Kolmogorov equation take for Markov chains?

8.5. It is possible for a process to satisfy the Chapmann–Kolmogorov equations and not be Markovian. For each $k = 0, 1, 2, \ldots$ let Y_k and Z_k be independent Bernoulli random variables, and independent for different k, with

$$P[Z_k = 1] = P[Y_k = 1] = 1/2,$$
$$P[Z_k = 0] = P[Y_k = 0] = 1/2$$

Let W_k be determined by

$$W_k = \begin{cases} 1, & Z_k = Y_k \\ 0, & Z_k \neq Y_k \end{cases}$$

Define, for $k = 0, 1, \ldots$,

$$X_{3k+i} = \begin{cases} W_k, & i = 0 \\ Y_k, & i = 1 \\ Z_k, & i = 2 \end{cases}$$

Show it satisfies the Chapmann–Kolmogorov equations, but is not Markov.

8.6. Consider the Markov chain which has the transition matrix

$$\mathbf{P} = \begin{bmatrix} 1 & 0 & 0 & 0 \\ 1/4 & 0 & 3/4 & 0 \\ 0 & 1/4 & 0 & 3/4 \\ 0 & 0 & 0 & 1 \end{bmatrix}$$

(a) Find the limiting distribution for this chain.

(b) Is this chain irreducible?

(c) Which states are transient, and which are recurrent?

(d) Find $P(T_1 = n|X_0 = 2)$.

(e) Find $P(T_4 = n|X_0 = 2)$.

8.7. Suppose that X_n is irreducible, aperiodic, and has a finite state space of N elements.

(a) Show that the real eigenvalues λ of the associated transition matrix are either

(b) If **P** is diagonalizable, conclude that it has limiting distribution

$$\mathbf{P} = \mathbf{C}^{-1}\mathbf{D}\mathbf{C}$$

where **D** is of the form

$$\mathbf{D} = \begin{bmatrix} 1 & 0 & 0 & \cdots & 0 \\ 0 & 0 & 0 & \cdots & 0 \\ 0 & 0 & 0 & \cdots & 0 \\ \vdots & \vdots & \vdots & \vdots & \vdots \\ 0 & 0 & 0 & \cdots & 0 \end{bmatrix}$$

8.8. For

$$\mathbf{P} = \begin{bmatrix} 1/2 & 1/2 & 0 \\ 0 & 1/4 & 3/4 \\ 3/4 & 1/4 & 0 \end{bmatrix}$$

(a) Find $\mathbf{P}^2$, $\mathbf{P}^3$, $\mathbf{P}^4$, $\mathbf{P}^n$.

(b) Show that the related Markov chain is irreducible and aperiodic.

(c) Find the stationary distribution.

(d) Does $\mathbf{P}^n$ converge?

8.9. Verify equation (8.6.2).

8.10. Set $n = 3$ in Section 8.7 and carry out the calculations explicitly.

8.11. Suppose that X_n is the symmetric random walk on the integers with an absorbing barrier at zero. Show that if $X_0 = k$, then $E[T_0]$ is infinite, for any value of k.

8.12. A gambler starts with N dollars, and bets on each toss of a fair coin, winning a dollar if heads comes up and losing a dollar if tails comes up. He stops as soon as he becomes broke, or as soon as he reaches his goal of $M > N$ dollars.

(a) What is the probability that he reaches his goal?

(b) Suppose that the coin is unfair, with $p = P[\text{heads}] \neq \frac{1}{2}$. What are his chances?

8.13. Show that the probability that a random walk with absorbing barriers reaches one of the barriers is 1.

8.14. Suppose that X_n is a Markov chain on the nonnegative integers, with

$$p_{ij} = P[X_{n+1} = j | X_n = i]$$

satisfying

$$p_{ij} = \begin{cases} s_i, & j = i - 1 \\ r_i, & j = i \\ b_i, & j = i + 1 \end{cases}$$

where $s_i + r_i + b_i = 1$, and $p_{00} = r_0 = 1$. This is called a birth and death chain.

(a) Suppose that $b_n = 0$, $s_n \neq 0$ and b_i, $s_i \neq 0$ for $i = 1, 2, \ldots, n - 1$.
   (i) Show that $0, 1, \ldots, n$ is an irreducible set of states.
   (ii) Find a formula for the stationary distribution π concentrated on $0, 1, \ldots, n$.

(b) Find a condition for a stationary distribution to exist, if b_i, $s_i \neq 0$ for all $i \geq 1$.

8.15. Consider a random walk with

$$P[X_{n+1} = i + 1 | X_n = i] = p$$

and

$$P[X_{n+1} = i - 1 | X_n = i] = 1 - p = q$$

(a) Show that

$$p_{00}(2k) = \binom{2k}{k} p^k q^k$$

(b) Use this, Stirling's formula, and the characterization of recurrence to show that return to zero is certain if $p = \frac{1}{2}$, and the probability of return is less than one if $p \neq \frac{1}{2}$.

(c) Verify that the probability of return is $1 - |p - q|$.

8.16. If π_1 and π_2 are two stationary distributions for a Markov chain, and $0 \leq a \leq 1$, show that $a\pi_1 + (1 - a)\pi_2$ is a stationary distribution.

8.17. Let X_n be the Markov chain with stochastic matrix

$$\mathbf{P} = \begin{bmatrix} 1 - \alpha & \alpha \\ \beta & 1 - \beta \end{bmatrix}$$

Find the stationary distribution for X_n, and show that it coincides with the limiting distribution.

8.18. **(a)** Find all the stationary distributions for

$$\mathbf{P} = \begin{bmatrix} 0 & 1 & 0 & 0 \\ 1/2 & 1/2 & 0 & 0 \\ 0 & 0 & 3/4 & 1/4 \\ 0 & 0 & 1/4 & 3/4 \end{bmatrix}$$

(b) Does $\mathbf{P}^n$ converge? If so, to what?

(c) Determine which states intercommunicate.

(d) Which states are transient and which are recurrent?

(e) Are all states aperiodic?

8.19. Show for a random walk on the integers that every state is recurrent, yet the mean recurrence time for every state is infinite.

8.20. Consider the Markov chain with transition matrix

$$\mathbf{P} = \begin{bmatrix} 0 & 1 & 0 \\ 1-p & 0 & p \\ 0 & 1 & 0 \end{bmatrix}$$

(a) Find P^2.

(b) Find P^n, $n \geq 1$.

8.21. Find any stationary distributions for the Markov chain with transition matrix

$$\mathbf{P} = \begin{bmatrix} 0.2 & 0.3 & 0.5 \\ 0.3 & 0.4 & 0.3 \\ 0.1 & 0.2 & 0.7 \end{bmatrix}$$

8.22. A matrix is called doubly stochastic if it is stochastic and each of its columns sum to 1.

(a) Show that the product of two stochastic matrices is stochastic.

(b) Show that the products of doubly stochastic matrices are doubly stochastic.

(c) Suppose that a doubly stochastic matrix has a limiting distribution. What must its form be?

(d) Show that an infinite doubly stochastic matrix cannot be positive recurrent.

8.23. Suppose that X_n is a branching process, where the family size Y is given by

$$P[Y = k] = 2^{-k+1}, \qquad k = 0, 1, 2, \ldots$$

and X_n be the size of the nth generation.

(a) Find the probability generating function in this case.

(b) Find the probability of ultimate extinction.

(c) Suppose that we start with five individuals; what is the probability of extinction?

8.24. Consider the transition matrix

$$
P = \begin{bmatrix}
0 & 1 & 0 & 0 & 0 \\
1 & 0 & 0 & 0 & 0 \\
0 & 0 & 0 & 0 & 1 \\
0 & 0 & 0 & 1 & 0 \\
0 & 0 & 1 & 0 & 0
\end{bmatrix}
$$

(a) Find the period of each of the states.

(b) Find P^n.

(c) Find all the stationary distributions for the related Markov chain.

8.25. Suppose that during each hour there is probability p that one customer will arrive at a store, and probability s that if there is a customer in the store, one customer leaves the store.

(a) Determine the transition function for the Markov chain X_n = the number of individuals in the store.

(b) Under what conditions is there a stationary distribution?

(c) If there is a stationary distribution, what is it?

8.26. Let

$$
P = \begin{bmatrix}
1/4 & 0 & 3/4 & 0 & 0 & 0 \\
0 & 1/2 & 0 & 1/2 & 0 & 0 \\
0 & 0 & 1/3 & 2/3 & 0 & 0 \\
0 & 0 & 1/2 & 1/2 & 0 & 0 \\
0 & 0 & 0 & 0 & 1 & 0 \\
1/2 & 0 & 0 & 0 & 0 & 1/2
\end{bmatrix}
$$

(a) Determine the transient and recurrent states.

(b) Suppose that $P[X_0 = S_5] = 1$. Find the expected length of time it takes to be absorbed into a set of recurrent states.

(c) Find the hitting probabilities f_{ij}.

Review of Linear Algebra

This appendix is intended as a self-contained collection of all the concepts from linear algebra that appear in the text. The primary focus is on developing results related to the diagonalization of matrices; however, some other material has been included for the sake of completeness.

A.1 MATRICES AND MATRIX ALGEBRA

An $m \times n$ **matrix A** is a rectangular array of numbers a_{ij},

$$
\mathbf{A} = \begin{bmatrix} a_{11} & a_{12} & \cdots & a_{1n} \\ a_{21} & a_{22} & \cdots & a_{2n} \\ \vdots & \vdots & \vdots & \vdots \\ a_{m1} & a_{m2} & \cdots & a_{mn} \end{bmatrix}
\tag{A.1.1}
$$

for which we introduce the shorthand $\mathbf{A} = (a_{ij})$. The numbers a_{ij} are called the **elements** or **entries** of $\mathbf{A}$.

Two matrices are said to be **equal** if they are the same size and their corresponding entries are equal,

$$
\mathbf{A} = \mathbf{B} \text{ iff } a_{ij} = b_{ij}
\tag{A.1.2}
$$

244

Matrices must be of the same dimensions to be added, and **addition** is elementwise

$$\mathbf{A} + \mathbf{B} = \mathbf{C} \text{ iff } a_{ij} + b_{ij} = c_{ij} \tag{A.1.3}$$

Scalar multiplication is defined similarly

$$k\mathbf{A} = \mathbf{C} \text{ iff } ka_{ij} = c_{ij} \tag{A.1.4}$$

where k is a real number. The **product** of two matrices $\mathbf{A}$ and $\mathbf{B}$ is defined if the number of columns of $\mathbf{A}$ is the same as the number of rows of $\mathbf{B}$, by

$$\mathbf{AB} = \mathbf{C} \text{ iff } c_{ij} = \sum_{k=1}^{n} a_{ik} b_{kj} \tag{A.1.5}$$

We note that even if $\mathbf{AB}$ and $\mathbf{BA}$ are both defined they are not usually equal. If $\mathbf{A} = (a_{ij})$ is a matrix, we define its **transpose, $\mathbf{A}^{\mathrm{T}}$**, to be the matrix obtained by interchanging the rows and columns of $\mathbf{A}$,

$$\mathbf{A}^{\mathrm{T}} = (a_{ji}) \tag{A.1.6}$$

We note $(\mathbf{AB})^{\mathrm{T}} = \mathbf{B}^{\mathrm{T}}\mathbf{A}^{\mathrm{T}}$ if either product is defined. A $1 \times n$ matrix

$$\mathbf{v} = (v_1, v_2, \ldots, v_n) \tag{A.1.7}$$

is called a row matrix, or **row vector,** and an $m \times 1$ matrix

$$\mathbf{w} = \begin{bmatrix} w_1 \\ w_2 \\ \vdots \\ w_m \end{bmatrix} \tag{A.1.8}$$

is called a column matrix, or **column vector,** and an $n \times n$ matrix is called a **square matrix.** If $\mathbf{w}$ and $\mathbf{y}$ are column vectors, we define the **inner product,** or dot product, of $\mathbf{w}$ and $\mathbf{y}$ as the number $\langle \mathbf{w}, \mathbf{y} \rangle$ defined by

$$\langle \mathbf{w}, \mathbf{y} \rangle = \mathbf{w}^{\mathrm{T}}\mathbf{y} = \sum_{i=1}^{n} w_i y_i \tag{A.1.9}$$

We shall write $\|\mathbf{w}\|$ for the length of $\mathbf{w}$, which is

$$\|\mathbf{w}\| = \sqrt{\langle \mathbf{w}, \mathbf{w} \rangle} \tag{A.1.10}$$

Vectors $\mathbf{w}$ and $\mathbf{y}$ are called **orthogonal** if

$$\langle \mathbf{w}, \mathbf{y} \rangle = 0 \tag{A.1.11}$$

Note that this corresponds to the notion of **perpendicularity.** The following notation shall be used for the **multiplicative** and **additive** identities:

$$
I_{n \times n} = \begin{bmatrix}
1 & 0 & 0 & \cdots & 0 \\
0 & 1 & 0 & \cdots & 0 \\
0 & 0 & 1 & \cdots & 0 \\
\vdots & \vdots & \vdots & \vdots & \vdots \\
0 & 0 & 0 & \cdots & 1
\end{bmatrix},
$$

(A.1.12)

$$
O_{n \times m} = \begin{bmatrix}
0 & 0 & 0 & \cdots & 0 \\
0 & 0 & 0 & \cdots & 0 \\
0 & 0 & 0 & \cdots & 0 \\
\vdots & \vdots & \vdots & \vdots & \vdots \\
0 & 0 & 0 & \cdots & 0
\end{bmatrix}
$$

The subscripts indicate the dimensions of the matrix, and will be omitted where they are obvious.

A collection of vectors which is closed under the operations of addition and scalar multiplication is called a **vector space.** A collection of vectors from a vector space, which themselves form a vector space, is called a **subspace** of the vector space. Let V be a vector space. A set of vectors $E = \{v_1, v_2, \ldots, v_k\}$ from V is said to be **linearly independent** if

$$
a_1 v_1 + a_2 v_2 + \cdots + a_k v_k = 0 \text{ iff } a_1 = a_2 = \cdots = a_k = 0 \quad \text{(A.1.13)}
$$

The **span** of $\{v_1, v_2, \ldots, v_k\}$ is the vector space $\mathbb{E}$ of all vectors which can be expressed as

$$
b_1 v_1 + b_2 v_2 + \cdots + b_k v_k \tag{A.1.14}
$$

where the b_i's are real numbers. If $\mathbb{E} = V$, the set is said to **span** V. A linearly independent set of vectors that spans a vector space is called a **basis** for the vector space, and every basis of a vector space will contain the same number of elements which is called the **dimension** of the vector space. The dimension of the span of the rows of a matrix A is called the **rank** of A. It is a nontrivial fact that the rank of A is the same as the rank of A^T for any matrix A.

A set of vectors $E = \{v_1, v_2, \ldots, v_k\}$ is called **orthonormal** if

$$
\begin{aligned}
&\text{(i) } <v_i, v_j> = 0, && i \neq j \\
&\text{(ii) } < v_i, v_i> = 1, && \text{for every } i
\end{aligned}
$$

(A.1.15)

The first condition is called **mutual orthogonality** and the second condition means the vectors are of unit length. A basis which is also an orthonormal set is called an **orthonormal basis.** Any basis $\{v_1, v_2, \ldots, v_k\}$ can be transformed into an orthonormal basis $\{w_1, w_2, \cdots w_k\}$ by the **Gram–Schmidt process:**

$$\mathbf{y}_1 = \mathbf{v}_1, \qquad\qquad\qquad \mathbf{w}_1 = \frac{\mathbf{y}_1}{\|\mathbf{y}_1\|}$$

$$\mathbf{y}_2 = \mathbf{v}_2 - <\mathbf{v}_2, \mathbf{w}_1> \mathbf{w}_1, \qquad\qquad \mathbf{w}_2 = \frac{\mathbf{y}_2}{\|\mathbf{y}_2\|}$$

$$\mathbf{y}_3 = \mathbf{v}_3 - <\mathbf{v}_3, \mathbf{w}_2> \mathbf{w}_2 - <\mathbf{v}_3, \mathbf{w}_1> \mathbf{w}_1, \qquad \mathbf{w}_3 = \frac{\mathbf{y}_3}{\|\mathbf{y}_3\|}$$

$$\vdots \qquad\qquad\qquad\qquad \vdots$$

$$\mathbf{y}_n = \mathbf{v}_n - <\mathbf{v}_n, \mathbf{w}_{n-1}> \mathbf{w}_{n-1} - \cdots - <\mathbf{v}_n, \mathbf{w}_1> \mathbf{w}_1, \qquad \mathbf{w}_n = \frac{\mathbf{y}_n}{\|\mathbf{y}_n\|}$$

A.2   SYSTEMS OF LINEAR EQUATIONS, LINEAR TRANSFORMATIONS

Matrices may be used to represent systems of linear equations, that is,

$$a_{11}x_1 + a_{12}x_2 + \cdots + a_{1m}x_m = y_1$$
$$a_{21}x_1 + a_{22}x_2 + \cdots + a_{2m}x_m = y_2$$
$$\vdots \qquad \vdots \qquad \vdots \qquad \vdots$$
$$a_{n1}x_1 + a_{n2}x_2 + \cdots + a_{nm}x_m = y_n$$

may be written as

$$\begin{bmatrix} a_{11} & a_{12} & \cdots & a_{1m} \\ a_{21} & a_{22} & \cdots & a_{2m} \\ \vdots & \vdots & \vdots & \vdots \\ a_{n1} & a_{n2} & \cdots & a_{nm} \end{bmatrix} \begin{bmatrix} x_1 \\ x_2 \\ \vdots \\ x_m \end{bmatrix} = \begin{bmatrix} y_1 \\ y_2 \\ \vdots \\ y_n \end{bmatrix}$$

or briefly,

$$\mathbf{A}\mathbf{x} = \mathbf{y} \qquad\qquad (A.2.1)$$

One can view this as a functional relation, which has the properties

$$\mathbf{A}(k\mathbf{x}) = k\mathbf{A}(\mathbf{x}) \qquad\qquad (A.2.2)$$

$$\mathbf{A}(\mathbf{x} + \mathbf{y}) = \mathbf{A}(\mathbf{x}) + \mathbf{A}(\mathbf{y}) \qquad\qquad (A.2.3)$$

and is called a linear transformation.

A consequence of the matrix formulation (A.2.1) of a linear system of equations is that solving a linear system is the same as solving a matrix equation. We assume the reader is familiar with the elimination method for solving (A.2.1), and recall that this method involves replacing **A** by a matrix which is obtained from **A** by a combination of the **elementary row operations:**

(i)   Interchange two rows of **A**.
(ii)   Multiply a row by a nonzero constant.
(iii)   Add to one row a multiple of another row.

A.3   SQUARE MATRICES

One of the most important quantities associated with a square matrix **A** is its **determinant,** which is denoted $|\mathbf{A}|$ and defined as follows:

(i)   If $\mathbf{A} = \begin{bmatrix} a & b \\ c & d \end{bmatrix}$, then

$$|\mathbf{A}| = ad - bc \qquad (A.3.1)$$

(ii)   For general **A** we define $|\mathbf{A}|$ by a reductive process,

$$|\mathbf{A}| = \sum (-1)^{i+j} a_{ij} |\mathbf{A}_{ij}| \qquad (A.3.2)$$

where $\mathbf{A}_{ij}$ is the matrix obtained by deletion of the ith row and the jth column from **A**. The sum is taken along any row or column of **A** and is independent of the choice of row or column you make for the summation. This method of calculating the determinant is called **expansion by cofactors** and each term $(-1)^{i+j} |\mathbf{A}_{ij}|$ in the sum is called the **cofactor** of a_{ij}.

Another way to define determinants is in terms of permutations. A **permutation** is a 1-to-1 function σ from $\{1, 2, \ldots, n\}$ into itself. We call σ a **transposition** if it interchanges two adjacent numbers:

$$\sigma(k) = \begin{cases} k, & k = 1, 2, \ldots, i-1, i+2, \ldots, n \\ i+1, & k = i \\ i, & k = i+1 \end{cases} \qquad (A.3.3)$$

Any permutation can be written as the composition of a finite number of transpositions

$$\sigma = \sigma_1 \circ \sigma_2 \circ \cdots \circ \sigma_m$$

Although there are many ways of expressing a permutation in terms of transpositions, the number of transpositions (m above) required is either even or odd for a given σ. We therefore can define the sign of a permutation by

$$\text{sign}(\sigma) = (-1)^m \qquad (A.3.4)$$

With this, the determinant of a square matrix $\mathbf{A}$ can be expressed as

$$|\mathbf{A}| = \sum_{\sigma} \text{sign}\,(\sigma) a_{1\sigma(1)} a_{2\sigma(2)} \cdots a_{n\sigma(n)} \tag{A.3.5}$$

We state without proof the important **multiplicative property of determinants,**

$$|\mathbf{AB}| = |\mathbf{A}||\mathbf{B}| \tag{A.3.6}$$

A square matrix $\mathbf{A}$ is said to be **invertible** if there exists a matrix, which is denoted $\mathbf{A}^{-1}$, such that

$$\mathbf{AA}^{-1} = \mathbf{A}^{-1}\mathbf{A} = \mathbf{I} \tag{A.3.7}$$

The matrix $\mathbf{A}^{-1}$ is called the **inverse** of $\mathbf{A}$. One may calculate the inverse of an invertible matrix $\mathbf{A}$ by using either an elimination technique, or by using determinants and cofactors. For elimination we write the augmented matrix

$$[\mathbf{A}|\mathbf{I}]$$

and by a sequence of elementary row operations transform this into

$$[\mathbf{I}|\mathbf{B}]$$

The matrix $\mathbf{B}$ will then be the inverse of $\mathbf{A}$. Denoting the entries of the inverse by a'_{ij}, they can be expressed in terms of determinants and cofactors by

$$a'_{ij} = \frac{1}{|\mathbf{A}|}\,((-1)^{i+j}|\mathbf{A}_{ji}|) \tag{A.3.8}$$

Inverses can be used to give the solutions to linear systems of equations, such as

$$\mathbf{Ax} = \mathbf{b}$$

provided $\mathbf{A}$ is invertible, by

$$\mathbf{x} = \mathbf{A}^{-1}\mathbf{Ax} = \mathbf{A}^{-1}\mathbf{b}$$

If you write this with the cofactor representation of the inverse, you obtain **Cramer's rule,**

$$x_i = \frac{|\mathbf{A}_i^{\mathbf{b}}|}{|\mathbf{A}|} \tag{A.3.9}$$

where $\mathbf{A}_i^{\mathbf{b}}$ is the matrix formed by replacing the ith column of $\mathbf{A}$ by $\mathbf{b}$.

We can connect some of these concepts by stating that the following are equivalent for an $n \times n$ matrix $\mathbf{A}$:

(i) $\mathbf{A}$ is invertible.
(ii) $|\mathbf{A}| \neq 0$
(iii) $\mathbf{A}\mathbf{x} = \mathbf{0}$ has only the trivial solution $\mathbf{x} = \mathbf{0}$.
(iv)   The rank of $\mathbf{A}$ is n.

Such a matrix is called **nonsingular,** otherwise the matrix is called **singular.**
    A square matrix $\mathbf{A} = (a_{ij})$ is called:

(i)   **symmetric** if $\mathbf{A}^\mathrm{T} = \mathbf{A}$.
(ii)   **antisymmetric** if $\mathbf{A}^\mathrm{T} = -\mathbf{A}$.
(iii)   **diagonal** if $a_{ij} = 0$ for $i \neq j$.
(iv)   **upper triangular** if $a_{ij} = 0$ for $i > j$.
(v)   **orthogonal** if it has orthonormal rows.

We note that:

(i) $|\mathbf{A}| = |\mathbf{A}^\mathrm{T}|$
(ii)   The determinant of an antisymmetric matrix is zero if n is odd.
(iii)   If $\mathbf{A}$ is diagonal $|\mathbf{A}|$ is the product of its diagonal elements.
(iv)   If $\mathbf{A}$ is upper triangular $|\mathbf{A}|$ is the product of its diagonal elements.
(v)   If $\mathbf{A}$ is orthogonal $\mathbf{A}^\mathrm{T}\mathbf{A} = \mathbf{I}$, and the determinant of $\mathbf{A}$ is ± 1.

Since for an invertible matrix $\mathbf{A}$, and a given matrix $\mathbf{B}$, if $\mathbf{B}\mathbf{A} = \mathbf{I}$,

$$\mathbf{B} = \mathbf{B}\mathbf{I} = \mathbf{B}(\mathbf{A}\mathbf{A}^{-1}) = (\mathbf{B}\mathbf{A})\mathbf{A}^{-1} = \mathbf{I}\mathbf{A}^{-1} = \mathbf{A}^{-1}$$

we conclude the following are equivalent for square matrices $\mathbf{A}$ and $\mathbf{B}$:

$$\mathbf{B}\mathbf{A} = \mathbf{I} \quad \text{iff} \quad \mathbf{A}\mathbf{B} = \mathbf{I} \quad \text{iff} \quad \mathbf{B} = \mathbf{A}^{-1}$$

Therefore, for orthogonal matrices $\mathbf{U}$

$$\mathbf{U}^\mathrm{T} = \mathbf{U}^{-1} \tag{A.3.10}$$

and $\mathbf{U}^\mathrm{T}$ is also orthogonal.

A.4   STRUCTURE OF SQUARE MATRICES

Two square matrices $\mathbf{A}$ and $\mathbf{B}$ are called **similar** if there exists an invertible matrix $\mathbf{R}$ such that

$$\mathbf{A} = \mathbf{R}^{-1}\mathbf{B}\mathbf{R} \tag{A.4.1}$$

which is in fact a symmetric definition since if $\mathbf{S} = \mathbf{R}^{-1}$, then

$$\mathbf{S}^{-1}\mathbf{A}\mathbf{S} = \mathbf{B} \tag{A.4.2}$$

A number λ is called an eigenvalue of the matrix $\mathbf{A}$ if

$$|\mathbf{A} - \lambda\mathbf{I}| = 0 \tag{A.4.3}$$

This is equivalent to the existence of nontrivial solutions to

$$\mathbf{A}\mathbf{x} = \lambda\mathbf{x} \tag{A.4.4}$$

which are called **right eigenvectors** for the matrix $\mathbf{A}$, and nontrivial solutions to

$$\mathbf{x}^{\mathrm{T}}\mathbf{A} = \lambda\mathbf{x}^{\mathrm{T}} \tag{A.4.5}$$

called **left eigenvectors** for the matrix $\mathbf{A}$. Thus $\mathbf{A}$ has zero as an eigenvalue iff $\mathbf{A}$ is singular. We also note that the eigenvalues of triangular or diagonal matrices are their diagonal entries.

The polynomial $|\mathbf{A} - \lambda\mathbf{I}|$ is called the characteristic polynomial of the matrix $\mathbf{A}$. A few of its properties and their consequences are:

(i) $|\mathbf{A} - \lambda\mathbf{I}|$ has degree n, leading coefficient 1, and constant term $(-1)^n|\mathbf{A}|$. Therefore, $\mathbf{A}$ has n complex eigenvalues, some of which may be repeated.

(ii) $|\mathbf{A}^{\mathrm{T}} - \lambda\mathbf{I}| = |\mathbf{A} - \lambda\mathbf{I}|$, so $\mathbf{A}$ and $\mathbf{A}^{\mathrm{T}}$ have the same eigenvalues.

(iii) $|\mathbf{A}^{-1} - \lambda\mathbf{I}| = (-\lambda)^n|\mathbf{A} - (1/\lambda)\mathbf{I}|/|\mathbf{A}|$, so the eigenvalues of $\mathbf{A}^{-1}$ are the reciprocals of the eigenvalues of $\mathbf{A}$.

(iv) $\mathbf{A}$ and $\mathbf{R}^{-1}\mathbf{A}\mathbf{R}$ have the same characteristic polynomial and hence similar matrices have the same eigenvalues. In general the eigenvectors are not the same.

We state without proof an interesting and nontrivial fact known as the **Cayley–Hamilton theorem:** If $f(x)$ is the characteristic polynomial of a matrix $\mathbf{A}$, then $f(\mathbf{A}) = \mathbf{0}$.

A matrix is said to be **diagonalizable** if it is similar to a diagonal matrix. To see how eigenvalues and eigenvectors are related to diagonalization, suppose that $\mathbf{A}$ is diagonalizable,

$$\mathbf{A} = \mathbf{R}^{-1}\mathbf{D}\mathbf{R} \tag{A.4.6}$$

and let e_i be given by

$$e_1 = (1, 0, 0, 0, \ldots, 0)^T$$

$$e_2 = (0, 1, 0, 0, \ldots, 0)^T$$

$$\vdots$$

$$e_n = (0, 0, 0, 0, \ldots, 1)^T$$

(A.4.7)

Write (A.4.6) in the form

$$RA = DR$$

so that

$$e_i^T RA = e_i^T DR$$

Noting that $e_i^T R$ is the ith row of R, and $e_i^T DR$ is the product of d_{ii} with the ith row of R, we conclude that the ith row of R is a left eigenvector of A with eigenvalue d_{ii}. By considering

$$AR^{-1}e_i = R^{-1}De_i$$

it follows that the ith column of R^{-1} is a right eigenvector of A with eigenvalue d_{ii}. From these we observe that the following are equivalent:

(i) A is diagonalizable.
(ii)   There exists an invertible matrix R whose rows are left eigenvectors of A.
(iii) There exists an invertible matrix R^{-1} whose columns are right eigenvectors of A.

A right eigenvector u, and a left eigenvector v^T, corresponding to different eigenvalues are orthogonal, since

$$\lambda_2 v^T u = (v^T A)u = v^T(Au) = v^T(\lambda_1 u) = \lambda_1 v^T u$$

Thus an $n \times n$ matrix A with n distinct eigenvalues λ_i, $i = 1, 2, 3, \ldots, n$, is diagonalizable,

$$A = R^{-1}DR$$

where R can be taken to be any matrix whose rows are the n left eigenvectors for A. R^{-1} will then be the matrix whose columns are the corresponding right eigenvectors for $\mathbf{A}$, suitably normalized so that the inner product of corresponding right and left eigenvectors is 1.

A.5 SYMMETRIC MATRICES

Any real symmetric matrix $\mathbf{A}$ has the property that all its eigenvalues are real. To show this, let λ (possibly complex) be an eigenvalue of $\mathbf{A}$, so that there exists $\mathbf{x}$ (possibly having complex entries), satisfying

$$\mathbf{A}\mathbf{x} = \lambda\mathbf{x}$$

Using

$$\mathbf{A}\bar{\mathbf{x}} = \bar{\lambda}\bar{\mathbf{x}}$$

and the fact that $\mathbf{A}$ is symmetric, it follows that

$$\lambda|\mathbf{x}|^2 = (\lambda\mathbf{x})^T\bar{x} = (\mathbf{A}\mathbf{x})^T\bar{\mathbf{x}} = \mathbf{x}^T(\mathbf{A}\bar{\mathbf{x}}) = \mathbf{x}^T(\bar{\lambda}\bar{\mathbf{x}}) = \bar{\lambda}|\mathbf{x}|^2$$

which shows λ is real.

An important result of Schur is that any real matrix $\mathbf{A}$, with all real eigenvalues, may be written $\mathbf{A} = \mathbf{R}^{-1}\mathbf{G}\mathbf{R}$, where $\mathbf{G}$ is upper triangular and $\mathbf{R}$ is orthogonal. We shall not prove this fact; however, it follows immediately that any real symmetric matrix is diagonalizable, since $\mathbf{R}^{-1} = \mathbf{R}^T$ implies that

$$\mathbf{G} = \mathbf{R}\mathbf{A}\mathbf{R}^T = \mathbf{R}\mathbf{A}^T\mathbf{R}^T = (\mathbf{R}\mathbf{A}\mathbf{R}^T)^T = \mathbf{G}^T$$

so $\mathbf{G}$ is diagonal. Since $\mathbf{R}$ is orthogonal, $\mathbf{A}$ is called **orthogonally diagonalizable.**

A.6 QUADRATIC FORMS

A symmetric matrix $\mathbf{A}$ defines what is called a quadratic form $Q(\mathbf{x})$ by

$$Q(\mathbf{x}) = \mathbf{x}^T\mathbf{A}\mathbf{x} = \sum\sum a_{ij}x_ix_j \tag{A.6.1}$$

A quadratic form is called positive definite if $Q(\mathbf{x}) \geq 0$, for any real $\mathbf{x}$, in which case $\mathbf{A}$ is called positive definite. Positive definiteness is equivalent to $\mathbf{A}$ having only positive eigenvalues. To show this, consider the orthogonal diagonalization of $\mathbf{A}$

$$\mathbf{A} = \mathbf{R}^{-1}\mathbf{D}\mathbf{R}$$

Since $\mathbf{R}$ is orthogonal $\mathbf{R}^{-1} = \mathbf{R}^T$, if $\mathbf{A}$ is positive definite we have

$$0 \leq \mathbf{x}^T\mathbf{A}\mathbf{x} = \mathbf{x}^T\mathbf{R}^T\mathbf{D}\mathbf{R}\mathbf{x}$$

Letting $\mathbf{x} = \mathbf{R}^T\mathbf{e}_i$, this becomes

$$0 \le d_{ii}$$

as required.

To show that this is sufficient, first note that $\mathbf{y}^T\mathbf{D}\mathbf{y} \ge 0$ for any $\mathbf{y}$, if $d_{ii} \ge 0$ for all i. Setting $\mathbf{y} = \mathbf{R}\mathbf{x}$ we have

$$0 \le \mathbf{y}^T\mathbf{D}\mathbf{y} = \mathbf{x}^T\mathbf{R}^T\mathbf{D}\mathbf{R}\mathbf{x} = \mathbf{x}^T\mathbf{A}\mathbf{x}$$

For 2×2 matrices $\mathbf{A}$ we can describe positive definiteness directly in terms of the entries of the matrix. The eigenvalues of

$$A = \begin{bmatrix} a & b \\ b & c \end{bmatrix}$$

are found from $|\mathbf{A} - \lambda\mathbf{I}| = 0$ or

$$(a - \lambda)(c - \lambda) - b^2 = 0$$

The two solutions

$$\lambda = \frac{(a + c) \pm \sqrt{(a + c)^2 - 4(ac - b^2)}}{2}$$

will be positive if and only if

$$a + c > 0 \text{ and } ac - b^2 > 0$$

or, equivalently,

$$a > 0, \qquad c > 0, \qquad ac - b^2 > 0 \qquad (A.6.2)$$

Index

255